I0031207

Berliner ökophysiologische
und phytomedizinische Schriften

Hrsg. von Christian Ulrichs und Carmen Büttner

Lebenswissenschaftliche Fakultät,
Humboldt-Universität zu Berlin

Band 46

Hrsg. von

Christian Ulrichs
Humboldt-Universität zu Berlin

und

Dr. Komi Fiaboe
International Centre of Insect Physiology and Ecology (*icipe*)

Development of entomopathogenic fungi as biopesticides for the management of Cowpea Aphid, *Aphis craccivora* Koch

Dissertation

Eingereicht an der

Lebenswissenschaftlichen Fakultat der

Humboldt-Universität zu Berlin

Von: Allan Ndua Mweke

4.11.1972, Nairobi, Kenya

Prasidentin

der Humboldt-Universität zu Berlin

Prof. Dr. Sabine Kunst

Dekan

der Lebenswissenschaftlichen Fakultät

Prof. Dr. Bernhard Grimm

Gutachter

1. Prof. Dr. Dr. Christian Ulrichs
2. Prof. Dr. Stefan Kühne

Bibliografische Information der Deutschen Nationalbibliothek
Die Deutsche Nationalbibliothek verzeichnet diese Publikation in der
Deutschen Nationalbibliografie; detaillierte bibliographische Daten sind im Internet
über http://dnb.d-nb.de abrufbar.
1. Aufl. - Göttingen: Cuvillier, 2018
    Zugl.: Berlin, Humboldt-Univ., Diss., 2018

© CUVILLIER VERLAG, Göttingen 2018
    Nonnenstieg 8, 37075 Göttingen
    Telefon: 0551-54724-0
    Telefax: 0551-54724-21
    www.cuvillier.de

    ISBN 978-3-7369-9908-4
    eISBN 978-3-7369-8908-5

Abstract

Cowpea (*Vigna unguiculata* L. Walp) is an important African indigenous vegetable in the tropics where it is considered as a key source of nutrients and income for small-holder farmers. The crop is drought tolerant and does well in poor soils because of its ability to fix atmospheric nitrogen. Cowpea is commonly used in soil fertility management and provides livestock fodder. However, the exploitation of the full potential of this crop is constrained by abiotic and biotic factors that lead to low yields in Africa. Among the biotic factor's cowpea aphid (*Aphis craccivora* Koch) is a major limiting factor in cowpea production and leads to yield losses of up to 100% when infestations occur in early stages of the crop and when control measures are delayed or not implemented. Where cowpea is grown as leafy vegetable, presence of honeydew produced by aphid on leaves renders them unfit for human consumption further occasioning yield losses. Chemical of insect pests' control is the most popular adopted management strategy but is associated with many negative effects on users, environment and beneficial organisms. This study evaluated pathogenicity of *Metarhizium anisopliae*, *Beauveria bassiana* and *Icaria* sp. isolates against *A. craccivora*. All the isolates were pathogenic to *A. craccivora* but *M. anisopliae* isolates ICIPE 62 and ICIPE 41, and B. bassiana ICIPE 644 outperformed all the other isolates by producing the highest mortalities of 90, 80 and 75% within the shortest time (LT_{50}) of 3.3, 3.6 and 3.7 days, respectively at conidial concentration of 1×10^8 conidia ml^{-1}. Evaluation of dose-dependent mortality was done for the three isolates and ICIPE 62 produced the lowest concentration that killed 50% of the tested insects ($LC_{50} = 2.3 \times 10^6$). Comparison of relative potency showed that ICIPE 62 was more potent than the other isolates and the same isolate produced the highest number of spores (4.5×10^7) on aphid cadavers 5 days post-treatment. In screenhouse evaluation oil and aqueous formulations of ICIPE 62 resulted in reduction of aphid population compared to control.

Field evaluation of aqueous and oil formulations did not show any significant effects on the aphids in the first season characterized by heavy and frequent rain and lower temperatures but were able to highly reduce aphid population in the second season which was dry with reduced and infrequent rainfall and elevated temperatures. However, application of Duduthrin in both seasons did not reduce aphid population compared to fungal treated plots. Treatment applications did not confer yield benefits in season 1(wet) but did in season 2 (dry) where EPF treated plots produced more leaf yield compared to other treatments though grain yield in season 2 was not different among the treatments.

Efficacy of combining intercropping of cowpea and maize and application of entomopathogenic fungi (EPF) was evaluated for three cropping seasons under field conditions. The performance of the EPF was compared to Duduthrin and untreated control in cowpea monocrop and cowpea-maize intercrop. In the first season which recorded high rainfall and cooler temperatures, the application of EPF and Duduthrin in either monocrop or intercrop did not reduce aphid populations. Similarly, in the second season with characteristic dry period with lower precipitation and higher temperatures, intercropping alone or application of Duduthrin in an intercrop did not reduce aphid population nor protect the crop from *A. craccivora* damage. However, at the same time, application of EPF in a maize-cowpea-intercrop reduced aphid population and protected the crop from aphid damage. The third season with average rainfall lower than season 1 but higher than season 2, application of EPF in the monocrop and intercrop significantly reduced aphid population. Comparison of leaf yield showed that EPF treated cowpea-maize intercrop did not produce higher yields than other intercrop treatment (maize-cowpea intercrop treated with Duduthrin and the untreated maize-cowpea intercrop) combinations in seasons 1 and 3. However, in season 2 (dry and hot) the cowpea leaf yield in cowpea-maize intercrop treated with EPF was similar to cowpea monocrop treated with EPF and higher than cowpea-maize intercrop treated with Duduthrin despite the monocrops having higher cowpea plant population. Cowpea grain yield in cowpea-maize intercrop was higher in season 2 among the intercrop treatment combinations. Application of Duduthrin did not increase leaf and grain yield production in all the three cropping seasons.

This study has identified ICIPE 62 as a potential fungal-based biopesticide for the management of cowpea aphid. The study also demonstrated the efficacy of combining intercropping and application of EPF as a viable control strategy for *A. craccivora* under field conditions. However, the success of this approach depends primarily on optimal weather conditions. Adoption of these control strategies could significantly reduce reliance on chemical pesticides and confer more benefits to small-holder vegetable producers.

Zusammenfassung

Die Kuhbohne (*Vigna unguiculata* L.Walp) ist ein wichtiges afrikanisches Gemüse in den Tropen. Hier stellt es eine wichtige Quelle für Nährstoffe und Einkommen für Kleinbauern dar. Die Pflanze ist dürreresistent und eignet sich gut für schlechte Böden, da sie in der Lage ist, Luftstickstoff zu binden. Kuhbohnen werden häufig im Rahmen des Bodenfruchtbarkeitsmanagements eingesetzt und dienen auch als Viehfutter. Die Ausschöpfung des vollen Potenzials dieser Kulturpflanze wird durch abiotische und biotische Faktoren eingeschränkt, die zu niedrigen Erträgen in Afrika führen. Unter den biotischen Faktoren ist die Schwarze Bohnenblattlaus (*Aphis craccivora* Koch) ein wesentlicher limitierender Faktor der zu Ertragsausfällen von bis zu 100 % führen kann. Dieses insbesondere, wenn der Befall in frühen Phasen der Kultur auftritt und wenn Bekämpfungsmaßnahmen verzögert oder nicht durchgeführt werden. Die chemische Bekämpfung von Schadinsekten ist die am weitesten verbreitete Strategie, wird aber mit ungewollten negativen Auswirkungen auf Anwender, Umwelt und Nutzorganismen verbunden.

In der vorliegenden Studie wurde die Pathogenität der entomopathogenen Pilze *Metharhizium anisopliae*, *Brassica bassiana* und *Iscaria* sp. gegen *A. craccivora* untersucht. Alle getesteten Isolate waren pathogen für A. craccivora. Dabei erzeugten die Isolate ICIPE 62, ICIPE 41 und ICIPE 644 innerhalb kürzester Zeit die höchste Mortalität (90, 80 und 75 %) (LT$_{50}$ 3.3, 3.6 und 3.7 Tage) bei einer konidialen Konzentration von 1×10^8 Konidien ml^{-1}. Die Bewertung der dosisabhängigen Mortalität erfolgte für die drei Isolate und ICIPE 62 zeigte sich bei mit der niedrigsten Konzentration wirksam. Zum Vergleich wurde der LC$_{50}$-Wert herangezogen. Der Vergleich der relativen Potenz zeigte, dass ICIPE 62 stärker war als die anderen Isolate und das gleiche Isolat produzierte die höchste Anzahl von Sporen auf Blattlauskadavern 6 Tage nach der Behandlung. Besonders unter kontrollierten Bedingungen im Gewächshaus führten öl- und wässrige-Formulierungen von ICIPE 62 im Vergleich zur Kontrolle zu einer reduzierten Blattlauspopulation und damit hohen Wirksamkeit.

Im Freilandversuch konnten wässrige und ölhaltige Formulierungen entomopathogener Pilze die Blattläuse in der ersten Saison nicht kontrollieren. Dieses hängt mit den herrschenden Witterungsbedingungen in der Saison zusammen. In der zweiten Saison wurde die Blattlauspopulation stark reduziert obwohl der Schädlingsdruck im Vergleich zur ersten Saison sehr hoch war. Die Anwendung von Duduthrin in beiden Anbauzeiträumen konnte die Blattlauspopulation nicht unter eine Schadschwelle senken. Die Behandlungen führten in der zweiten Anbausaison zu einer signifikanten Erhöhung der Ernteerträge.

Die Bekämpfung von *A. craccivora* durch einen Zwischenfruchtanbau (Kuhbohnen und Mais) sowie die Anwendung von entomopathogenen Pilzen wurde über drei Anbauperioden unter Praxisbedingungen evaluiert. Dabei wurden zur Vergleichbarkeit Behandlungen mit dem chemischen Pflanzenschutzmittel Duduthrin sowie eine unbehandelte Kontrolle herangezogen. In der ersten Saison konnte die Anwendung von entomopathogenen Pilzen als auch von Duduthrin in der Monokultur und in der Bohnen-Mais-Zwischenkultur keine Blattläuse kontrollieren. In der zweiten Anbausaison konnten der Zwischenfruchtanbau allein und auch die Anwendung von Duduthrin die Blattlauspopulation ebenfalls nicht senken. Im Gegensatz dazu waren der Einsatz von entomopathogenen Pilzen in einem Anbausystem von Mais und Kuhbohne erfolgreich in der Bekämpfung des Schädlings. Der Vergleich der Blatterträge zeigte, dass mit Pilzen behandelte Kuhbohnen im Anbau mit der Zwischenkultur Mais in den Anbauzeiträumen 1 und 3 keine höheren Erträge erzielten, im Vergleich zur Behandlung von Mais mit Duduthrin bzw. der unbehandelten Kontrolle. In der zweiten Anbausaison erzielte die Behandlung von Kuhbohnen mit Pilzen im Zwischenanbau bzw. in der Bohnenmonokultur hingegen gute Ergebnisse. Die alleinige Anwendung von Duduthrin brachte in allen drei Jahreszeiten keinen Ertragsvorteil bei Blatt und Saatgut. Die vorliegende Studie hat ICIPE 62 als potenzielles Isolat für das Management von *A. craccivora* identifiziert. Die Studie zeigte auch die Wirksamkeit der Kombination von Zwischenfruchtanbau und Anwendung von entomopathogenen Pilzen als praktikable Bekämpfungsstrategie für *A. craccivora* unter Feldbedingungen. Ein Erfolg dieses Ansatzes ist primär von optimalen Witterungsbedingungen abhängig. Eine Akzeptanz dieser Bekämpfungsstrategien durch Farmer kann den Einsatz chemischer Pestizide verringern und die ungewollten Nebenwirkungen des Pflanzenschutzmitteleinsatzes verringern.

Table of Contents

LIST OF TABLES

LIST OF FIGURES

LIST OF ABBREVIATIONS

AIVs   African Indigenous Vegetables
ANOVA   Analysis of Variance
BMBF   German Federal Ministry of Education and Research
BMZ   German Federal Ministry of Economic Cooperation and Development
DAE   Days after Emergence
DRIP   Dissertation Research Internship Program
EPF   Entomopathogenic fungi
FAO   Food and Agriculture Organization of the UN
FL   Fiducial Limit
GLM   Generalized Linear Model
Ha   Hectare
HCD   Horticultural Crops Directorate
ICIPE   International Centre of Insect Physiology and Ecology
IPM   Integrated Pest Management
KEPHIS   Kenya Plant Health Inspectorate Service
LC_{50} Lethal Concentration that kills 50% of tested population
LT_{50} Lethal Time (time taken to kill 50% of tested population)
PDA   Potato Dextrose Agar
Rpm   Revolutions per Minute
SIDA   Swedish International Development Cooperation Agency
SDC   Swiss Agency for Development and Cooperation
SSA   Sub-Saharan Africa
Spp   Species (plural)
Tukey's HSD   Tukey's Honestly Significant Difference
UV   Ultra Violet

1.0 GENERAL INTRODUCTION

1.1 Background

Cowpea, *Vigna unguiculata* (L.) Walp (Fabaceae) is an important leguminous crop in semi-arid tropics and is well adapted to drought (Akibonde and Maredia, 2011; Horn et al., 2016; Yadav et al., 2017).

Estimated annual production of cowpea is 6.2 million metric tons (MT) occupying 14.5 million hectares of land in more than 45 countries across the world (Abate et al., 2012). Africa is the leading producer of cowpea with Nigeria, Niger and Burkina Faso accounting for 80% of the world's production (FAO, 2016). Kenya produces about 2% of the total world production mainly in the drier regions (HCD, 2014; FAO, 2016).

Consumption of cowpea either as leaves, immature pods, green peas and dry grains provides local populations affordable sources of proteins, essential vitamins and minerals (Ghaly and Alkoaik, 2010; Hall, 2012). Cowpea is used mainly as grain legume in West Africa and as a leafy vegetable in East Africa (Chiulele et al., 2011; Hall, 2012; Rusike et al., 2013). Leafy cowpea is an important African indigenous vegetable (AIV) in Kenya that is widely grown and consumed in urban and rural areas (Abukutsa-Onyango, 2010; Rusike et al., 2013). Cowpea also has medicinal benefits as its consumption has been shown to reduce environmental enteric dysfunction (Trehan et al., 2015).

The vegetative parts of cowpea plant are fed to livestock as fodder (haulms) and farmers earn income from selling them in the dry season (Rusike et al., 2013). The spreading cowpea varieties protect the soil against soil erosion and help in weed control (Garko et al., 2016). The crop improves soil fertility and soil structure by providing organic matter from crop residues and fixing atmospheric nitrogen into the soil through symbiosis with nodule forming bacteria such as *Bradyrhizobium* spp. (Mucheru-Muna, 2010; Schipanski and Drinkwater, 2012; Dwivedi et al., 2015).

1.2 Problem statement and justification of the study

In Africa, yields of cowpea are low, ranging between 100 and 250 kg/ha (Omongo et al., 1997; Baidoo et al., 2012) compared to potential yields of 3000 kg/ha in United States of America (Rusoke and Rubaihayo, 1994; Hall, 2012). Arthropod pests and diseases as well as of use of inferior varieties and/or farmer saved seeds and poor soil fertility are the key limiting factors in cowpea production (Obopile, 2006; Dugje et al., 2009). Cowpea aphid, *Aphis craccivora* Koch (Aphididae), is a major pest of cowpea that attacks the crop in all the stages of its growth (Blackman and Easton 2006; Kusi et al., 2010; Souleymane et al., 2013). The pest contributes to yield losses of up to 100% through disruption of plant physiological growth

11

by sacking the plant sap and slowing of photosynthetic process by reducing leaf surface area exposed to light by depositing honeydew on the leaves (Sorensen, 2009). Additional yield losses are also due to transmission of more than 30 plant viruses including Cowpea aphid-borne mosaic virus (CABMV) (Blackman and Eastop, 2000; Smith and Boyko 2007; Damiri et al., 2013).

Management of the cowpea aphid is mainly based on use of chemical pesticides, which have been reported to be expensive and ineffective (Hassan, 2013). Furthermore, pesticides have many undesirable impacts like development of resistance due to indiscriminate use, human health concerns due to exposure during application and chemical residues on plant products, bioaccumulation in the environment and disruption of ecological services of beneficial organisms (Sánchez-Bayo, 2011; Baidoo et al., 2012; El-Heneidy et al., 2015). The negative environmental impacts of the synthetic pesticides have accelerated search for alternative crop protection products leading to increased development of less toxic compounds based on naturally occurring toxins from micro-organisms (Mazid *et al.*, 2011).

Several integrated management approaches targeting cowpea aphid have been developed and used with varying degrees of success (Afun et al., 1991; Egho, 2010). These strategies have employed use of resistant varieties (Huynh et al., 2013; Smith and Chuang, 2014), use of biological control agents including the use of entomopathogenic fungi (EPF) formulated as mycoinsecticides (Hajek and Delalibera, 2010), manipulation of agro ecosystems like intercropping (Hassan, 2013), and targeted insecticide application as opposed to routine and blanket spraying (Egho and Enujeke, 2012). The plant sap sucking soft bodied aphids are susceptible to attack by EPF under natural environment and epizootics due to infection by EPF have been reported (Pell et al., 2003; Roy et al., 2010). The EPF belong to the Zygomycetes and Hyphomycetes groups but the Zygomycetes class form the larger group of EPF attacking aphids including *A. craccivora* (Humber 1991). Several isolates of *Metarhizium anisopliae* (Metschnikoff) Sorokin, *Beauveria bassiana* (Balsamo) Vuillemin, *Lecanicillium* spp and *Isaria* spp are commercial products currently available for use in management of several aphid species. For example, *B. bassiana* (registered as BotaniGard® and Naturalis-L®), *M. anisopliae* (registered as Met52®), *Isaria javanica* (Frieder and Bally), (Samson and Hywel-Jones), (registered as Preferal®) and *Lecanicillium* spp. (registered as Vertalec®) are all used for aphid control in Europe and North America (Zimmerman, 1992; Cook et al., 1996; Whipps, 1997; Fravel et al., 1998; Wraight and Carruthers, 1999; Copping and Menn, 2000; Hynes and Boyetchko, 2006; Jandricic et al., 2014).

Fungal based biopesticides are being used as commercial crop protection products for management of aphids in Asian, Latin American as well as and European countries which account for the greatest market share of these products with Africa registering and using the lowest percentage (Faria and Wraight, 2007). Biopesticides are increasingly becoming viable

12

alternatives for control of insect pests due to their safety to users, non-target beneficial arthropods and environment and their compatibility with IPM strategies.

Though there are previous studies on pathogenicity of *M. anisopliae*, *B. bassiana* and *Isaria* spp on *A. craccivora* and other aphid species (Ekesi et al., 2000, Sahayaraj and Borgio 2010, Saranya et al., 2010; Bayisse et al., 2016), the studies did not evaluate the promising isolate in this study; ICIPE 62 against *A. craccivora*. Additionally, ICIPE 62 has been used to control other aphid species in vegetables and is commercially available as a biopesticide in Kenya. Moreover, none of the isolates evaluated in this study have been tested for pathogenicity against *A. craccivora* before. In Kenya, there is an EPF based biopesticide (Met 62®-*M. anisopliae* isolate ICIPE 62) developed by the International Centre of Insect Physiology and Ecology (*icipe*) in collaboration with Real IPM Kenya against major vegetable arthropod pests including aphid species (http://www.realipm.com/). However, this product does not include *A. craccivora* as one of the target pests hence this study evaluated the pathogenicity of this isolate among others for potential development of a biopesticide that can be used in the management of this pest.

1.3 General objective

This study aimed at developing and optimizing entomopathogenic fungi as biopesticides for the management of *A. craccivora* within the context of cowpea IPM.

1.4 Specific objectives

1. Screen entomopathogenic fungi isolates for their virulence against the cowpea aphid (*A. craccivora*) and select candidate isolates that can be developed into a biopesticide and used in an IPM system
2. Evaluate different formulations of the selected isolate for the management of *A. craccivora* on cowpea under field conditions
3. Assess the efficacy of intercropping cowpea with maize and application of selected EPF isolate for the management of *A. craccivora*.

1.5 Research questions

This study set out to answer the following questions:
4. Does pathogenicity of entomopathogenic fungi to *Aphis craccivora* vary within and among isolates of different species?
5. Is production of fungal spore on insect cadavers positively related to isolate(s) virulence?
6. Does formulation type influence performance of entomopathogenic isolates both in screenhouse and field conditions?

13

7. Is combination of intercropping cowpea and maize and application of entomopathogenic fungi more effective in suppressing aphid population under field conditions compared to application of Duduthrin or non-application of either EPF or Duduthrin in a cowpea maize intercrop?

1.6 References

Abate T, Alene AD, Bergvinson D, Shiferaw B, Silim S, Orr A and Asfaw S. 2012. Tropical grain legumes in Africa and South Asia: knowledge and opportunities. International Crops Research Institute for the Semi-Arid Tropics, Nairobi, pp 112.

Abukutsa MOO. 2010. African indigenous vegetables in Kenya: Strategic repositioning in the horticultural sector. Inaugural Lecture, Jomo Kenyatta University of Agriculture and Technology, Nairobi, Kenya. 30 April.

Afun JVK, Jackai LEN and Hodgson CJ. 1991. Calendar and monitored insecticide application for the control of cowpea pests. *Crop Protection*, 10 (5):363-370.

Akibode S and Maredia MK. 2011. Global and regional trends in production, trade and consumption of food legume crops. Department of Agricultural, Food, and Resource Economics Michigan State University. Staff Paper Series, 2012-10.

Baidoo PK, Baidoe-Ansah D and I. Agbonu I. 2012. Effects of Neem (*Azadirachta indica* A. Juss). Products on *Aphis craccivora* and its predator *Harmonia axyridis* on cowpea. *American Journal of Experimental Agriculture*, 2:198-206.

Bayissa W, Ekesi S, Mohamed SA, Kaaya GP, Wagacha JM, Hanna R and Maniania NK. 2016. Selection of fungal isolates for virulence against three aphid pest species of crucifers and okra. *Journal of Pest Science*, 90:355-68.

Blackman RL and Eastop VF. 2000. Aphids on the world's crops: an identification and information guide (No. Ed. 2). John Wiley & Sons Ltd. 476.

Blackman RL and Eastop VF. 2006. Aphids on the world's herbaceous plants and shrubs. Chichester, UK: John Wiley & Sons Ltd, pp. 1460.

Chiulele RM, Mwangi G, Tongoona P, Ehlers JD and Ndeve AD. 2011. Assessment of farmers' perceptions and preferences of cowpea in Mozambique. *African Crop Science Conference Proceedings*, 10:311-318.

Cook RJ, Bruckart WL, Coulson JR, Goettel MS, Humber RA, Lumsden RD, Maddox JV, McManus ML, Moore L, Meyer SF and Quimby Jr PC. 1996. Safety of microorganisms intended for pest and plant disease control: A framework for scientific evaluation, *Biological Control*, 7:333–351.

Copping LG and Menn JJ. 2000. Biopesticides: a review of their action, applications and efficacy. *Pest Management Science*, 56 (8): 651-676.

Damiri BV, Al-Shahwan IM, Al-Saleh MA, Abdalla OA and AmerMA. 2013. Identification and characterization of cowpea aphid-borne mosaic virus isolates in Saudi Arabia. *Journal of Plant Pathology*, 79-85.

Dugje IY, Omoigui LO, Ekeleme F, Kamara AY and Ajeigbe H. 2009. Farmers' guide to cowpea production in West Africa. IITA, Ibadan, Nigeria.

Dwivedi SL, Sahrawat K, Upadhyaya H, Mengoni A, Galardini M, Bazzicalupo M, Biondi EG, Hungria M, Kaschuk G, Blair MW and Ortiz R. 2015. Advances in host plant and rhizobium genomics to enhance symbiotic nitrogen fixation in legumes, in: D.L. Sparks (Ed.), Advances in Agronomy, Elsevier Inc., Amsterdam, pp. 1–116.

Egho EO. 2010. Comparative studies on insect species of cowpea (*Vigna unguiculata*) (L) Walp) in two agro-ecological zones during the early cropping season in Delta State, Southern Nigeria. *Agriculture and Biology Journal of North America*, 1 (5): 946-949.

Egho EO and Enujeke EC. 2012. Minimizing insecticide application in the control of insect pests of cowpea (*Vigna Unguiculata* (L) WALP) in Delta state, Nigeria. *Sustainable Agriculture Research*, 1 (1): 87.

Ekesi S, Akpa AD, Onu I, Ogunlan MO. 2000. Entomopathogenicity of *Beauveria bassiana* and *Metarhizium anisopliae* to the cowpea aphid, *Aphis craccivora. Archives of Phytopathology and Plant Protection*, 33:171–180.

El-Heneidy AH, Khidr AA and Taman AA. 2015. Side-effects of insecticides on non-target organisms: 1-In Egyptian Cotton Fields. *Egyptian Journal of Biological Pest Control*, 25 (3):685.

Faria MR and Wraight SP. 2007. Mycoinsecticides and mycoacaricides: a comprehensive list with worldwide coverage and international classification of formulation types. *Biological Control*, 43 (2):237-256.

FAO, 2016. Food and Agriculture Organization of the United Nations, Rome, Italy. http://faostat.fao.org/default.aspx.

Fravel DR, Connick Jr W. and Lewis JA. 1998. Formulation of microorganisms to control plant diseases. In: Formulation of microbial biopesticides, Springer, Dordrecht, pp. 187-202.

Garko MS, Mohammed IB and Fulari MS. 2016. Performance of cowpea [*Vigna unguiculata* (l.) Walp.] varieties as influenced by weed control treatments in the Sudan Savanna of Nigeria. *International Journal of Scientific and Research Publications*, 6:135-140.

Ghaly AE and Alkoaik FN. 2010. Extraction of protein from common plant leaves for use as human food. *American Journal of Applied Sciences*, 7 (3):331.

Hajek AE and Delalibera I. 2010. Fungal pathogens as classical biological control agents against arthropods. *Bio Control*. 55 (1):147-158.

Hall AE. 2012. Phenotyping cowpea for adaptation to drought. *Frontiers in Physiology*, 3, 1-8.

Hassan S. 2013. Effect of variety and intercropping on two major cowpea (*Vigna unguiculata* Walp) field pests in Mubi, Adamawa state, Nigeria. *International Journal of Agricultural Research and Development*, 1:108-109.

Horn LN, Ghebrehiwot HM and Shimelis HA. 2016. Selection of novel cowpea genotypes derived through gamma irradiation. *Frontiers in Plant Science*, 7:262.

Horticultural Crops Directorate (HCD). 2014. Annual report.

Humber RA. 1991. Fungal pathogens of aphids. In: Peters DC, Webster JA, Chlouber CS (eds). Aphid-plant interactions: Populations to molecules, 45–56 Agricultural Experiment Station, Division of Agriculture, Oklahoma State University, Stillwater.

Huynh BL, Ehlers JD, Close TJ, Cisse´ N, Drabo I, Boukar O, Lucas MR, Wanamaker S, Pottorff M and Roberts PA. 2013. Enabling tools for modern breeding of cowpea for biotic stress resistance Translational genomics for crop breeding. Wiley, London, pp. 183–199.

Hynes RK and Boyetchko SM. 2006. Research initiatives in the art and science of biopesticide formulations. *Soil Biology and Biochemistry*, 38 (4):845-849.

Jandricic SE, Filotas M, Sanderson JP and Wraight SP. 2014. Pathogenicity of conidia-based preparations of entomopathogenic fungi against the greenhouse pest aphids *Myzus persicae*, *Aphis gossypii*, and *Aulacorthum solani* (Hemiptera: Aphididae). *Journal of Invertebrate Pathology*, 118: .34-46.

Kusi F, Obeng-Ofori D, Asante SK and Padi FK. 2010. New sources of resistance in cowpea to the cowpea aphid (*Aphis craccivora* Koch) (Homoptera: Aphididae). *Journal of the Ghana Science Association*, 12 (2): 95-104.

Mazid S, Kalida JC and Rajkhowa RC. 2011. A review on the use of biopesticides in insect pest management. *International Journal of Science and Advanced Technology*, 1 (7):169-178.

Mucheru-Muna M, Pypers P, Mugendi D, Kung'u J, Mugwe J, Merckx R and Vanlauwe B. 2010. A staggered maize–legume intercrop arrangement robustly increases crop yields and economic returns in the highlands of Central Kenya. *Field Crops Research*, 115 (2):132-139.

Obopile M. 2006. Economic threshold and injury levels for control of cowpea aphid, *Aphis craccivora* Linnaeus (Homoptera: Aphididae) on cowpea. *African Plant Protection*, 12:111–115.

Omongo CA, Ogenga-Latigo MW, Kyamanywa S and Adipala E. 1997. The effect of seasons and cropping systems on the occurrence of cowpea pests in Uganda. In *African Crop Science Conference Proceedings*, 3:1111-1116.

Pell JK, Eilenberg J, Hajek AE and Steinkraus DC. 2001. Biology, ecology and pest management potential of Entomophthorales. In: Butt TM, Jackson C, Magan N (eds) Fungi as biocontrol agents: progress, problems and potential. CAB International, Wallingford, pp. 71–153.

Roy HE, Brodie EL, Chandler D, Goettel MS, Pell JK, Wajnberg E and Vega F. 2010. Deep space and hidden depths: understanding the evolution and ecology of fungal entomopathogens. *Biological Control*, 55:1–6.

Rusike JG, van den Brand S, Boahen K, Dashiell S, Kantengwa J, Ongoma DM, Mongane G, Kasongo ZB, Jamagani R, Aidoo R and Abaidoo R. 2013. Value chain analyses of grain legumes in Kenya, Rwanda, Eastern DRC, Ghana, Nigeria, Mozambique, Malawi and Zimbabwe. (No. 1.2. 6, 1.3. 4). N2Africa.

Rusoke DG and Rubaihayo PR. 1994. The influence of some crop protection management practices on yield stability of cowpea. *African Crop Science Journal*, 2:43-48.

Sahayaraj K and Borgio JF. 2010. Virulence of entomopathogenic fungus *Metarhizium anisopliae* (Metsch.) Sorokin on seven insect pests. *Indian Journal of Agricultural Science*, 44:195–200.

Sánchez-Bayo F. 2011. Impacts of agricultural pesticides on terrestrial ecosystems. In: Sánchez-Bayo, F, van den Brink, PJ, Mann, R (eds.) Ecological Impacts of Toxic Chemicals. Bentham Science Publishers, Online, pp. 63-87.

Saranya SR, Ushakumari S, Jacob S and Philip BM. 2010. Efficacy of different entomopathogenic fungi against cowpea aphid, *Aphis craccivora* (Koch). *Journal of Biopesticides*, 3:1138–142.

Schipanski ME and Drinkwater LE. 2012. Nitrogen fixation in annual and perennial legume-grass mixtures across a fertility gradient. *Plant and Soil*, 357 (1-2):147-159.

Smith CM and Boyko EV. 2007. The molecular bases of plant resistance and defense responses to aphid feeding: current status. *Entomologia Experimentalis et Applicata*, 122 (1):1-16.

Smith CM and Chuang WP. 2014. Plant resistance to aphid feeding: behavioral, physiological, genetic and molecular cues regulate aphid host selection and feeding. *Pest Management Science*, 70:528–540.

Souleymane A, Aken'Ova ME, Fatokun CA and Alabi OY. 2013. Screening for resistance to cowpea aphid (*Aphis craccivora* Koch) in wild and cultivated cowpea (*Vigna unguiculata* Walp.) accessions. *International Journal of Science, Environment and Technology*, 2 (4): 611 – 621.

Trehan I, Benzoni NS, Wang AZ, Bollinger LB, Ngoma TN, Chimimba UK, Stephenson KB, Agapova SE, Maleta KM and Manary MJ. 2015. Common beans and cowpeas as complementary foods to reduce environmental enteric dysfunction and stunting in Malawian children: study protocol for two randomized controlled trials. *Trials*, 16 (1):520.

Whipps JM. 1997. Developments in the biological control of soil-borne plant pathogens. *Advances in Botanical Research*, 26:1–134.

Wraight SP and Carruthers RL. 1999. Production, delivery, and use of mycoinsecticides for control of insect pests of field crops. In: Hall FR, Menn JJ, (eds). Methods in Biotechnology, Vol. 5: Biopesticides: Use and Delivery. Totowa, New Jersey, USA: Humana Press, pp. 233-69.

Yadav T, Nisha KC, Chopra NK, Yadav MR, Kumar R, Rathore DK, Soni PG, Makarana G, Tamta A, Kushwah M and Ram H. 2017. Weed management in cowpea-A Review. *International Journal of Current Microbiology and Applied Sciences* 6 (2):1373-1385.

Zimmermann G. 2007 Review on safety of the entomopathogenic fungus *Metarhizium anisopliae*. *Biocontrol Science and Technology*, 17 (9):879-920.

2.0 SCIENTIFIC BACKGROUND

2.1 *Vigna unguiculata* L. Walp

Origin: Cowpea (*Vigna unguiculata* L. Walp.) is an annual drought tolerant legume that is adapted to different soil types and different cropping systems whose origin has been traced to Southern Africa region (Singh et al., 1997). The West African countries including Nigeria, Niger, Burkina Faso, Benin, Togo and Cameroon have the biggest diversity of cultivated cowpea (Ng and Marechal; 1985; Ng, 1995; Padulosi and Ng, 1997; Timko and Singh, 2008).

Global production: Cowpea (*Vigna unguiculata* L. Walp).is grown mostly in tropical Africa (Ofuya, 1997) and Africa produces the largest proportion of the world's production with Nigeria, Niger and Burkina Faso producing 80% of world's production (Mortimore et al., 1997; FAO, 2016). In 2016, the world production was 6,991,174 metric tons with a total value of US$ 18 billion while the total area under production was 12.3 million hectares. In the same year Africa produced 6,739,689 metric tons representing 96% of the world production (FAO, 2016).

Production in Kenya: Major cowpea production localities in Kenya are largely in the arid and semi-arid areas since it is a drought tolerant crop and farmers can harvest even when cereal crops like maize and sorghum fail (Saidi et al. 2010). In 2016, the total area under production was 31,020 hectares producing 115,801tons valued at Kenyan shillings 2.4 billion representing about 2% of the world production with Makueni County producing 35% of the total production (HCD, 2016). Areas of production include Machakos, Kitui, Makueni, Tharaka, Mbeere, Taita Taveta, Kwale, Kilifi, Lamu, Kisii, Migori, Homabay, Siaya, Kisumu and Bugoma The cowpea varieties grown by farmers can be categorize into 4 according to their seed colour or mode of growth (1) the cream types which have cream colored seeds, (2) the crowder type with black spots or brown colour, (3) the black eye types whose seeds are white with black eye and (4) other types with intermediate colours. Some commercial varieties in Kenya and East African region include Kunde M66, Ken Kunde 1(KK1) Ken Kunde 3(KK3), K80, KVU 27-31 and KVU 419.

Uses: Cowpea crop has a wide adaptability to different climatic conditions and it is cultivated in warm regions of the world mainly for its edible seed, however, the crop is also an important source of vegetable and, it is one of the most important African leafy vegetables (Hall, 2012; Rusike et al., 2013). Cowpea does well in poor soils because it has tolerance to low soil fertility and it has ability to fix atmospheric nitrogen in association with root nodule forming bacteria (*Bradyrhizobium* spp (Schipanski and Drinkwater LE 2012; Ddamulira et al., 2015) and it also able tolerate a wide range of soil pH (Mucheru-Muna, 2010).

In Kenya and other East African countries, cowpea has wide nutritional and agronomic uses. Young leaves are used as a vegetable while the seeds and young pods constitute a rich source of dietary protein. Dried shoots and roots are used as fuel (Hall, 2012; Trehan et al., 2015). Cowpea seed is a nutritious component in the human diet as it contains about 25% protein and 64% carbohydrate, and 2% fat (Akibode, 2011; Owolabi et al., 2012).The leaves have higher protein content compared to seed (Baker et al. 1989; Nielsen et al., 1993). Cowpea is widely consumed all over the world, mainly in rural populations, and satisfy a considerable proportion of the protein requirements (Oiye et al., 2009; Ghaly and Alkoaik, 2010).

Above ground parts of cowpea plant excluding the pods are used as fodder for livestock (haulms) and a source of income for farmers who harvest and sell during dry period (Singh et al., 1997). The spreading and indeterminate or semi-determinate varieties of cowpea provide ground cover against soil erosion while at the same time suppressing weeds (Singh et al., 1997; Mucheru-Muna, 2010). The cowpea crop residues when ploughed into the soil provide organic matter that improves soil fertility (Mucheru-Muna, 2010). Cowpea being a leguminous crop fixes atmospheric nitrogen into the soil through symbiosis with nodule forming bacteria (*Bradyrhizobium* spp), (Singh et al., 1997; Schipanski and Drinkwater, 2012; Dwivedi et al., 2015).

2.2 The Cowpea aphid *Aphis craccivora* (Koch)

Taxonomic description: Compared to other aphid species, *A. craccivora* is a relatively small aphid. Apterous females are characterized by black or dark brown body, brown to yellow legs and a prominent cauda. Nymphs are waxy compared to adults. Adults are distinguished from other closely related aphid species by the presence of 6-segmented antennae with black distal part of femur, siphunculi and cauda (Blackman and Eastop, 2006). Winged (alate) females of *A. craccivora* have distinct dorsal cross bars on the abdomen (Blackman and Eastop, 2000).

Distribution and occurrence: *Aphis craccivora* is widely distributed in the world and it has been reported in regions/countries where it was absent mainly due to changing climate but it is more endemic in the tropics (Blackman and Eastop, 2000).

Biology and ecology

Aphis craccivora has a wide distribution in the tropics where females reproduce parthenogenetically but sexual morphs have been reported in temperate regions (Blackman and Eastop, 2007). Cowpea aphid females are ovoviviparous and retain their eggs inside their bodies and give birth to nymphs. Small colonies of *A. craccivora* establish on actively growing plant parts like leaves, tips and young stems and are frequently found in association with ants (Flatt and Weisser, 2000; Espadaler et al., 2012). A number of biotypes of *A. craccivora* have identified (Ofuya, 1997; Sorensen, 2009).

Development of *A. craccivora* is influenced by climatic conditions including temperature (24-28.5°C), relative humidity 65% RH, hours of sunshine (day length-L: D 16:8) and precipitation (Mayeux, 1984). Host plant biochemistry like low levels of hydrocarbon positively influences development of alate individuals (Mayeux, 1984). The lifespan of adult's ranges between 5-15 days and under favourable weather conditions *A. craccivora* completes a generation in 10 to 20 days. An adult aphid can produce 20 nymphs in one day and developmental period between first instar and adult is between 3-5 days while a single adult female can produce up to 100 nymphs in their life time (Ofuya, 1997). Weather conditions, soil moisture content and fertility as well as host plant status influence growth, development, reproduction and the lifespan of *A. craccivora* (Ofuya 1997).

Host range

Aphis craccivora is a highly polyphagous aphid species feeding on Leguminoseae group of plants including cowpea (*Vigna unguicalata* (L.) (Walp.), groundnut (*Arachis hypogaea* L.), mungbean (*Vigna radiata* (L.) R. Wilczek), pigeonpea (*Cajanus cajan* (L.) Huth), chickpea (*Cicer arietinum* L.), green beans (*Vicia* spp. and *Phaseolus* spp.), lupins (*Lupinus angustifolius* L.), lentil (*Lens esculenta*) and lucerne (*Medicago sativa* L.). It is also reported as a minor pest on other leguminous non-leguminous crops, such as cotton and citrus (Blackman and Eastop, 2006; Brady and White, 2013).

Economic importance of *Aphis craccivora*

Direct crop damage: *Aphis craccivora* causes direct damage by sucking plant sap and injecting toxins into the phloem during all plant growth stages including seedlings, flowers and pods and this damage is by both adults and nymphs (Ofuya, 1997; Huynh et al., 2015). When heavy infestations occur during early plant growth stages young plants wither and eventually die and those that survive are characterized by stunted growth, distorted leaves and experience delayed flowering and lower yields (Ofuya, 1995). Heavy infestation by cowpea aphid at podding stage can reduce seed yield (Ofuya, 1997).

Indirect damage: High population of *A. craccivora* produces high amounts of honeydew on plant leaf surfaces thereby promoting growth of the sooty mold fungus which interferes with respiratory and photosynthesis capacity of the plant by altering biochemical and physiological processes of infested plants thereby reducing plant growth and associated yield (Gomez et al., 2004; Sorensen, 2009; Goławska et al., 2010). Honeydew also reduces quality of cowpea leaves and renders them inedible thus contributing to yield loss. *Aphis craccivora* is a known vector of more 30 plant viruses including cowpea mosaic virus, ground nut rosette virus (GRV), subterranean clover stunt virus (SCSV) (clover stunt virus), Bean common mosaic virus (BCMV) (bean mosaic virus, bean western mosaic virus, mungbean mosaic virus) (Atiri et al., 1986; Blackman and Eastop, 2000; Brault et al., 2010).

2.3 Management strategies

Chemical control

Small scale and subsistence farmers in Africa rely heavily on the use of chemicals to control cowpea aphid mainly because aphids are susceptible to most insecticides and also because it is virtually impossible to produce cowpea profitably without use of pesticides (Waddington et al., 2010; Egho and Enujeke, 2012). Different chemical groups have demonstrated their efficacy against *A. craccivora* and have been used in its management. These chemicals include synthetic pyrethroids, pyrethrins, organophosphates, carbamates and insect growth regulators (Maienfisch et al., 2001; Liu et al., 2011; Radha, 2013; Gowtham et al., 2016). However, use of synthetic chemicals is becoming increasingly untenable due to the environment pollution and contamination, high costs, safety of the users and residues in produce as well as development of resistance by the target pests (Sorensen, 2009; Dewhirst, 2010; Ferreira et al., 2013). Use of pesticides on leafy cowpea can significantly reduce yield due to observation of post harvest intervals as cowpea leaves are harvested regularly and the commonly used pesticides have longer post harvest intervals. There is also the health risk on consumers as the farmers may not observe the safe harvesting intervals especially when the demand is high (Mweke et al., 2016).

Cultural control

A number of cultural practices have been used in the management of *A. craccivora* in the tropics. Timing of planting to coincide with low aphid population pressure has been recommended (Jackai, 1985; Egho, 2010). Closely spacing cowpea and intercropping with cereals like maize, sorghum and millet has been shown to reduce cowpea aphid infestation (A'Brook, 1968; Hassan, 2013). The intercrop is known to hinder aphid movement by creating a barrier between the cowpea and aphids (Ezueh, 1991; Trenbath, 1993; Nampala 2002). The increased density from intercropping also creates a micro-climate within the crop canopy thus disrupting visual search of the host crop by the pest and may attract predators (Trenbath, 1993; Nampala et al., 1999). Dense sowing and early planting or delayed planting (late planting) weed control, crop rotation and intercropping are some of the cultural practices successfully used in the management of *A. craccivora* (Mayeux 1984; Nampala et al., 2002; Rizk, 2011). Densely sown crops create pseudo resistance thus reducing infestation by the pests (Farrel, 1976; Ofuya, 1989). Planting at the onset of rains reduces infestation by *A. craccivora* on cowpea (Jackai, 1985). Changing planting time creates asynchrony between crop phenology and insect pest (Ferro, 1987) and pest infestation may not coincide with the most vulnerable crop growth stage which in turn reduces or delays pest establishment on the crop (Dent, 1991).

Use of plant resistance

Plant improvement through breeding of resistant cultivars is a promising alternative strategy for aphid control in cowpea (Huynh et al., 2013; Smith and Chuang, 2014). Use of cowpea

21

varieties resistant to *A. craccivora* has been employed in management of the pest across the world (Jackai and Singh, 1988; Ofuya 1988a). The cowpea resistance to aphids is controlled by a mono dominant gene and has been attributed mainly to antibiosis as well as presence of phenols and or flavonoids in those varieties (Ofuya, 1988b; Lattanzio et al., 2000; Smith and Clement, 2012; Huynh et al., 2015). However, most of the resistance is more effective at seedling than at podding stage (Singh et al., 1990; Ofuya, 1993; Kamphuis et al., 2012). Resistance to *A. craccivora* has been reported to break down when the varieties are exposed to *A. craccivora* population from different regions from which the varieties were bred and tested, hence use of varietal resistance to *A. craccivora* alone is not a sustainable management approach for the aphid (Messina et al., 1985, Ofuya, 1997). There are some wild *Vigna* accessions with known resistance *A. craccivora* (Singh et al., 1990) but efforts to cross the cultivated *Vigna* species with wild *Vigna* species have been unsuccessful (Ng, 1995). Attempts have also been made to cross insect resistant *Vigna* species and cultivated *Vigna unguiculata* through biotechnology (Murdock, 1992) but this approach is prone to many challenges because insects may develop resistance to the transgenic cowpea since it is based on allelochemicals governed by a single gene and the allelochemicals produce toxins that may harm beneficial natural enemies of *A. craccivora* (van Emden, 1991; Smith and Clement, 2012). Plant genetic engineering has been successfully used in controlling plant viruses and may therefore be applied to control cowpea virus diseases transmitted by the cowpea aphid (Boxtel et al., 2000; Citadin et al., 2011).

Biological control

Predators

Use of natural enemies to control cowpea pests has been well documented (Ofuya and Akingbohungbe, 1988; Singh et al., 1990) and this has been more successful in the tropics (Gullan and Cranson, 1994). The Coccinellidae, Syrphidae, Ceccidomyiidae, Chrysopidae and Anthocoridae are the major predators of aphids (Chaudhary and Singh, 2012). The adult and larvae of lady beetles, lacewing larvae and syrphid fly larvae are the most common predators feeding on aphid under natural environment (Völkl, et al., 2007). Effective aphid control by natural enemies is more pronounced in controlled environments as compared to field conditions since rapid population buildup of aphids has been observed even in the presence of natural enemies (Singh et al., 1990, Ofuya, 1991). One of the approaches used to enhance effective control of aphids by natural enemies is periodic release of the predators in field, however, this is not applicable in tropical Africa since there are no mass producers of the natural enemies in major cowpea growing areas and even if there were, poor resource farmers may not afford them (Ofuya, 1995). The best alternative to this approach is conservation approach through judicious use of pesticides to avoid natural enemies' mortality (Ofuya, 1997).

22

Parasitoids

Aphids are reported hosts of more than 600 species of parasitoids of the hymenopteran group (Hymenoptera: Braconidae) (Stary, 2013). Aphid parasitoids in the subfamily Aphidiinae, are the most abundant and specialist aphid parasitoids and are the only endoparasitoids attacking aphids (Stary, 1970). These parasitoids regulate aphid population under natural environment and they have been used in biological control of aphids (Kambhampati *et al.*, 2000; Powell and Pickett, 2003; Boivin et al., 2012). *Aphis craccivora* is a known host of different species of aphidiines including *Aphidius colemani* Viereck, *Lysiphlebus fabarum* (Marshall), *L. confuses* Tremblay and Eady, and *L testaceipes* Cresson (Ofuya, 1995; Boivin et al., 2012). The Aphidiinae has more than 400 species with worldwide distribution (Stary´ 1988, Smith and Kambhampati 2000). The *Aphidius*, *Praon*, *Diaeretiella*, *Trioxys* and *Ephedrus* genera comprise species that are commonly used in biological control (Wei et al., 2005, Vollhardt et al., 2008). Some commercial products have been developed and are being used for management of aphids in developed world (Dassonville et al., 2012), however such products are rare in Africa.

Use of entomopathogens in aphid control

Bacteria, entomophagous nematodes, viruses, entomopathoegnic fungi, and protozoa are the major entomopathogens of arthropods and have been used as biological control agents for many years (Butt, 2000). Entomopathogenic fungi that are parasites of arthropods are valuable biocontrol agents and are compatible with integrated pest management (Shah and Pell, 2003; Lopes et al., 2011). Entomopathogenic fungi are a diverse group consisting of approximately 1000 species reported from many taxonomical divisions of the fungal kingdom (Kaya and Vega 2012). Susceptibility of aphids to EPF is well documented and epizootics of fungal diseases of aphids under natural environment have been reported (Milner, 1997; Shah and Pell 2003; Roy et al., 2010). For example, Ekesi et al., 2000; demonstrated potential of *M. aniopliae* and *B. bassiana* in management of *A. craccivora* under laboratory conditions. Among the fungal pathogens of aphids Zygomycetes and Hyphomycetes groups form the majority, however, entomophthora ean fungi belonging to the class Zygomycetes are the major pathogens of aphics including *A. craccivora* (Humber 1991). Important genera of EPF used in biological control include *Beauveria, Metarhizium* and *Lecanicillium*. These fungi infect insect through contact and penetrates the cuticle and sporulates in the haemocoel using nutrients from the insect causing eventual death through production of toxins and are suitable when targeting sucking pests. (Samson et al., 1988; Glare and Milner, 1991; Shah and Pell, 2003). Entomopathogenic fungi are applied in the form of conidia or mycelium using conventional pesticides application equipments and germinate on insect body after application to initiate infection (Mazid et al., 2012). Currently, various strains of EPF including *Lecanicillium* sp., *B. bassiana, Metarhizium anisopliae, Paecilomyces* sp. and *Nomuraea rileyi* are being used in biological control of aphids (Vu et al., 2007; Kim et al., 2008; Selvaraj et al., 2010). Entomptahogenic fungi belonging to the hyphomycete genus *Metarhizium* have

23

been isolated from infected insects and soil around the world (Roddam and Rath, 1997). Some isolates of this fungus have a narrow host range, but the group infects a wide spectrum of arthropod pests in many orders (Roberts and St. Leger, 2004). *Metarhizium anisopliae* (Metschnikoff) Sorokin is easy to mass produce and has been used for insect-pest control for more than a century (Roberts and St. Leger, 2004; Jaronski, 2013).

Beauveria bassiana has a worldwide distribution and is commonly found in natural environment and these properties have made it a good target for use in insect pest control (Boucias and Pendland, 2012; Ferron, 1978; Ferron et al., 1991; Vega and Blackwell, 2005). *Beauveria* produces different secondary metabolites including beauvericin, bassianolides, beauveriolides and oxalic acid among others and these metabolites are involved in its pathogenecity and virulence (Xiao et al., 2012).

Integrated management approach

Integrated pest management strategies are sustainable approaches to pest problems and are meant to reduce pest damage and maximize yield while minimizing undesirable impacts of pest control practices. Integrated management of cowpea pests including *A. craccivora* has been employed with desirable outcomes. For example, monitored pesticide applications based on established pest damage threshold has been shown to protect cowpea against *A. craccivora* and produce similar yields to routine insecticide application and hence conferring farmers with the benefits of reduced production cost and minimal pesticide residues (Afun et al., 1991; Egho and Enujeke 2012). Cultural practices such as intercropping cowpea with cereals and monitored pesticide applications have been proven to be more effective and profitable in controlling *A. craccivora* than weekly application of insecticides (Nabirye et al., 2003). Manipulation of planting dates in combination with plant density and insecticide applications resulted in increased cowpea yield and reduced cost of pest management (Karungi et al., 2000).

Use of certified cowpea seed cultivars, planting time and reduced use of insecticides is a proven and effective pest management technique of cowpea insect pests and has been shown to reduce pest damage, increased yield and has the advantage of reducing environmental pollution (Asante et al., 2001).

The potential for use of insect pathogens for management of crop pests was demonstrated by Augustino Bassi in 1835 when he described the "green muscardine" disease caused by *B. bassiana* in silkworm (Steinhaus, 1956; Faria and Wraight, 2007). Since then a lot of attention has been focused on development of commercial product based on entomopathogens for pest control driven partly by safety concerns associated with synthetic pesticides (Chandler et al., 2011; Jaronski, 2013). To date there are more than 100 commercial products that have been developed from entomopathogenic fungi and are being used in pest management (Butt *et al.*, 2001; Faria and Wraight, 2007; Roy et al., 2010; Mazid et al., 2012). These products are mainly based on *Beauveria, Metarhizium, Paecilomyces (Isaria)* and *Verticillium* (Faria and Wraight, 2007).

In evaluating successful use of EPF in pest management it is important to consider many aspects and accruing benefits since direct comparisons with chemical insecticides is misleading (Shah and Pell, 2003). In deciding on use of EPFs in IPM, it is necessary to consider technical efficacy in combination with practical efficacy (cost and adoption by users), commercial viability (cost benefit analysis), sustainability (long term control) and/or public benefit (safety), (Gelernter and Lomer 2000; Zimmermann, 2007). Safety of any pest control product to human, environment and non-target organisms is an essential criterion for evaluation and acceptability of the product and EPFs have been proven to meet this criterion and are therefore safer alternative in IPM systems compared to chemical pesticides (Goettel and Hajek 2000; Pell et al., 2001). Successful use of EPFs also requires identification of unique and specific markets for the products such as protected environments like greenhouses (Gelernter and Lomer 2000). Entomopathogenic fungi are best suited in situations where immediate pest eradication is not essential and where pest populations are to be maintained below economic threshold and some crop damage is permissible. These EPFs form an essential part of IPM when used in combination with other pest management strategies (Shah and Pell, 2003).

Current trends in use of biopesticides in pest management

Biopesticide is defined as mass-produced agent manufactured from a living microorganism or a natural product and sold for the control of plant pests (Chandler et al., 2011). The microbial based biopesticides are also referred to as biocontrol agents (BCAs). Environmental concerns associated with the use of synthetic pesticides and compatibility of biopesticides with integrated pest management strategies coupled with increased demand for reduction on use of pesticides by consumers and rising demand for organic products have led to increased research on the potential of biopesticides to replace pesticides in pest management (Pickett et al., 1995; Copping and Menn, 2000; Chandler et al., 2011). Unlike synthetic pesticides that have a similar mode of action against pests (neurotoxicity), biopesticides control pests in a myriad of actions including disrupting mating processes, induce anti-feedant activities, suffocate and predispose the insects to desiccation (Olson, 2015). These biopesticides are also target specific, not harmful to the environment and users and do not disrupt ecological balance between pests and beneficial organisms and have short re-entry intervals after application and leave no residues on the produce (Lacey and Siegel, 2001; Olson, 2015, Mweke et al., 2016).

Despite the huge potential that exist for the biopesticides their adoption is hampered by lack of information by potential users on their availability and effectiveness, low efficacy under field conditions, short shelf life, high cost of production and a small market niche, regulatory restrictions regarding registration as well as health and ecological concerns (Chandler et al.,

25

2011; Glare et al., 2012; Arora et al., 2016). These challenges are however, being addressed through research and development of new techniques using molecular biology and biotechnology that are geared towards improving production, application techniques as well as development of broad-spectrum products with improved efficiency, and increased product shelf life and overall efficacy (Butt et al., 1999; Chandler et al., 2011).

The market share for biopesticides has grown from 2.9% in 2006 to 10% in 2010 with a total value of US$ 1 billion representing 4.2% of the total pesticides market share globally (Leng et al., 2014). The 2016 market share was valued at US$ 2.83 Billion and is expected to reach a market share of 15.43% by 2022 (https://www.marketsandmarkets.com). The current market growth rate for biopesticides is estimated to be between 16 and 20% compared to 3% growth for synthetic pesticides annually and is projected to hit capitalization to the tune of US$ 10 billion by 2017 (Marrone, 2007; Leng et al., 2014; Olson et al., 2015). Africa has the lowest consumption of biopesticides in the world and the market was estimated to be worth $23 million 2003 (Guillon, 2003). In Kenya in 2002, biopesticides market share was $1.15 million (2%) of the total $57.4 million total pesticides market. (Wabule et al., 2004). This growth is being favoured by regulatory restrictions on the registration of new synthetic pesticide molecules, high cost of developing and registering chemical pesticides, safety of biopesticides as well as growing demand for organic products that are free from pesticide residues (Leng et al., 2014; Mishra et al., 2015; Olson, 2015).

2.4 Biotic factors-host and pathogen interactions that affect performance of entompathogens

Utilization of fungal pathogens as biologcal control agents in pest management has involved studies to understand their biology and ecology as well as the mechanisms involved in host recogntion, host reaction after infection and host defense mechanisms (Shahid et al., 2012; Ortiz-Urquiza and Keyhani, 2013). Pathogen host intaraction can be summarized as follows: (i) adhension of the infective propagules to the host (ii) host recognition and enzyme production (iii) germination of the propagule on the insect cuticle (iv) penetration of the host cuticle (v) sporulation in the host haemocoel (vi) production of toxins leading to death of the host and (vii) production of infective propagules on the insect cadaver (Shahid et al., 2012). Succcessful infection is preceded by attachment of the conidia to the host body (adhehnsion) and is regulated by proteins (Adhesins) and is facilitated by production of enzymes that degrade host cuticle to facilitate penetration (Boucias et al., 1998; Cosentino-Gomes et al., 2013).

Attack by entomopathogens elicits defense respose on the part of the arthropod pests that may include behavioural, stractural or environmental adaptation. Some insect raise body temperatures to levels unfavurable to development of infective propagules (thermo-regulation) while others lower body temperatures to enhance their immunity. Alteration of metabolic activities, non responsiveness to semiochemicals and change in sexual activites are also common response mechanisms employed by infected hosts (Inglis et al., 1996; Roy et al., 2006; Hunt et al., 2011; Hunt et al., 2016).

Successful infection process by the entomopathogens on the host involves evasion or overcoming of the arthropod immune system by the EPF (Vega at al., 2012) This is done through production of proteins (enzymes) such hydrophobins, adhesins and secondary metabolites that degrade the insect cuticle and overcome the immune system and antimicrobial activities of the host (Ortiz-Urquiza and Keyhani, 2013). To complete infection process some *Metarhizium* strains release toxins e.g. destruxin in the insect haemolymph (Amiri-Besheli et al., 2000) and these result in rapid death of their hosts compared to those starins that do not produce destruxin and kill their host slowly (Samuels et al., 1988).

2.5 Abiotic factors affetcing performnce ofentomopathogens
The efficacy and viability of EPF as biocontrol agents is adversely affected by prevailing environmental conditions like sunlight. rainfall, temperature, and humidity because EPFs are highly sensitive to environmental cond tions (Roy et al., 2006; Jaronski 2009; Eyheraguibel et al., 2010).

Solar radiation
Ultraviolet radiation is a major limiting factor on the efficacy of EPF propagules under natural environment as spore viability deteriorates upon exposure to UV light due to induced morphological and genetic changes through denaturation of EPF DNA (Rangel et al., 2008; Rodrigues et al., 2016). UV radiaton also adversely affects germination of conidia and *B. bassiana* is known to lose its infectivity after exposure to sunlight for a few hours (Bell, 1974; Costa et al., 2012). Physiological state of the infective spores determines their interaction with solar radiation as it has been demonstrated that actively growing conidia are more susceptible to UV light that resting spores (Braga et al., 2001). Reaction to UV light also varies among species, strains within species and species within groups (Fargues et al., 1996; Rodrigues *et al.*, 2016). Incorporaton of UV protectants in formulations of EPF not only reverses negative impacts of the UV light but also improves photochemical properties of the infective propagules (Maniania et al , 1993; Cohen et al., 2001; Rodrigues et al., 2016). Oil formulations of EPF are more stable when exposed to solar radiation compared to aqueous formulations (Alves et al., 1998; Inglis et al., 2002).

Relative humidity
High relative humidity (> 90%) is required for infection process and in production of highly infective conidia on insect cadavers after application of EPF and subsequently low atmospheric moisture adversely affects conidia germination and by extension performance of EPF (Ferron et al., 1991; Hajek and St. Leger, 1994). However, infection at low relative

humidity has been demonstrated (Ramoska, 1984; James et al, 1998). Manipulation of crop habitat through irrigation improves relative humidity and enhances pathogen infection as well efficacy of EPFs in insect pest management (Bateman et al., 1993; Hajek and St. Leger, 1994; Brooks et al., 2004). Formulating EPF as oil or oil emulsions as opposed to aqueous improves infectivity at low relative humidity (Inglis et al. 2002).

Temperature

Temperature is one of the abiotic factors that greatly affect performance of EPF in outdoor applications (Roy et al., 2006). It affects conidial germination and subsequent mycelial growth thus negatively influencing infection and virulence as well as survival and sporulation (Vidal et al., 1997; Inglis et al., 1997; Yeo et al., 2003; Nussenbaum et al., 2013). EPF in the genera Hyphomycetes perform optimally (i.e. germinate, grow and sporulate) within a temperature range of 20 and 30°C (Rangel et al., 2010). Beyond 37°C growth of EPF is severely hampered (Inglis et al., 2001; Chandra Teja and Rahman, 2016), however the *Metarhizium* species have highest reported thermo-tolerance with temperatures between 35 and 40 °C permitting conidial germination and mycelial growth and sporulation (Thomas and Jenkins, 1997; Milner et al., 1997; Dimbi et al., 2004). Physiological processes that occur during fungal germination and growth (disappearance of trehalose in conidia and change of fatty acids composition of the cell membrane) reduce heat tolerance (Thevelein, 1984; Pupin et al., 2000). Some *Metarhizium* isolates have proven ability to recover, germinate, grow and sporulate even after prolonged exposure to high temperatures and such isolates are suitable for development of biopesticides for use in the tropics where ambient air temperatures remain high throughout the year and where pests have many generations in a year (Rangel et al., 2010). Strains of EPF that can grow and produce infective spores within a wide temperature range are suitable for use as biopesticides mainly in arid and semi-arid regions (Chandra Teja and Rahman, 2016). Optimum temperature range for EPF vary across regions and hence it is critical to evaluate potential EPFs for development and commercialization based on the target market in terms of prevailing environmental conditions (Santos et al., 2011).

Rainfall

Heavy rainfall after application of EPF reduces its efficacy by washing away a considerable amount of infective propagules coming into contact with the target insect (Inglis et al., 2000; Inglis et al., 2002). To reduce the negative impacts of rain on EPFs stickers are used during formulation or spray application as most stickers are compatible with oil and oil-emulsion formulations of EPFs (Bernhard et al., 1998). Oil formulations of EPF have been shown to reduce the impact of rain on the EPF by increasing retention of the EPF on the leaf surface (Wraight and Ramos 2002). Addition of stickers to the oil formulations improves retention of EPF propagules on the leaf surface and thus increases chances of infection (Inglis et al., 2002). However, some stickers may interfere with the transfer of infective fungal propagules to the insect body and therefore they should be chosen carefully, and preference should be given to those that remain in liquid form after application (Inglis et al., 2002).

Formulation of entomopathogenic fungi

Formulation of EPF and the mode of application significantly affect their efficacy in controlling pests under field conditions (Inglis et al., 2002). This is so primarily because entomopathogenic fungi infect their host through the cuticle and hence contact of the infective spores of the fungi with the host is crucial (Inglis et al., 2002). The commonly used EPF formulations include solid state formulations like baits and encapsulate and liquid formulations mainly aqueous, oiland oil emulsions. Adjuvants, protectants and desiccants that prolong storage period and enhance efficacy and protect against UV and reduce desiccation of the spores are important components of the EPF formulations (Wraight et al. 2001; Inglis et al., 2002; Reddy et al., 2008). The infective spores (asexual spores) of most EPFs including *Beauveria* and *Metarhizium* have hydrophobic cell walls containing glycoproteins called hydrophobins and this makes them suitable for use in oil formulation (Inglis et al., 2002). Spray oils which are derivatives of petroleum products are commonly used as oil formulations for EPF because they are easily available and are affordable, have no toxic effects on the EPF spores and produce uniform viscosity across a range of temperatures (Burges, 1998). Oil formulations are preferred to aqueous formulations because conidia is uniformly suspended, improve storage time of the EPF without losing infectivity and produces smaller droplet sizes that enhances uniform coverage over long distances, protect against UV radiation, prevent desiccation, enhance infection under low atmospheric relative humidity.Oil based formulations also spreads infective spore more uniformly, enhance attachment and penetration of the insect cuticle and the oils are more stable (low volatility) and reduce application volumes (bulkiness) (Alves et al., 1998; Wraight et al., 2001; Inglis et al., 2002; Guinossi et al., 2012).

Formulation of hydrophobic asexual spores in aqueous solution has many limitations however, surfactants are added to improve suspension of the infective propagules but with detrimental outcomes on the viability of the spores (Jaronski, 1997; Bernhard et al., 1998; Wraight and Carruthers, 1999).

Desirable characteristics of commercial mycoinsecticides products include economical and easy to produce, longer storage stability, persistence residual activity, ease of handling, constitution and application as well as consistent performance against target pests under different environmental conditions (Lacey et al., 2001).

2.6 References

Abate T, Alene AD, Bergvinson D, Shiferaw B, Silim S, Orr A and Asfaw S. 2012. Tropical grain legumes in Africa and South Asia: knowledge and opportunities. International Crops Research Institute for the Semi-Arid Tropics, 112 pp.

A'Brook J. 1968. The effect of plant spacing on the numbers of aphids trapped over the groundnut crop. *Annals of Applied Biology*, 61: 289-294.

Adati T, Tamo` M, Yusuf SR, Downham MCA, Singh BB and Hammond W. 2008. 'Integrated pest management for cowpea-cereal cropping systems in the West African savannah'. *International Journal of Tropical Insect Science*, 27:123–137.

Afun JVK, Jackai LEN and Hodgson CJ. 1991. Calendar and monitored insecticide application for the control of cowpea pests. *Crop Protection*, 10:363-370.

Akibode S and Maredia MK. 2011. Global and regional trends in production, trade and consumption of food legume crops. Department of Agricultural, Food, and Resource Economics, Michigan State University.

Alves RT, Bateman RP, Prior C and Leather SR. 1998. Effects of simulated solar radiation on conidial germination of *Metarhizium anisopliae* in different formulations. *Crop Protection*, 17:675-79.

Amiri-Besheli B, Khambay B, Cameron S, Deadman ML and Butt TM. 2000. Inter-and intra-specific variation in destruxin production by insect pathogenic *Metarhizium* spp., and its significance to pathogenesis. *Mycological Research*, 104 (4):447-452.

Arora NK, Verma M, Prakash J and Mishra J. 2016. Regulation of biopesticides: Global concerns and policies. In Bioformulations: for Sustainable Agriculture. Springer, New Delhi, 283-299 pp.

Asante SK, Tamo M and Jackal LEN. 2001. Integrated management of cowpea insect pests using elite cultivars, date of planting and minimum insecticide application. *African Crop Science Journal*, 9:655-665.

Atiri GI and Thottappilly G. 1985. *Aphis craccivora* settling behaviour and acquisition of cowpea aphid-borne mosaic virus in aphid-resistant cowpea lines. *Entomologia Experimentalis et Applicata*, 39 (3):241-245.

Atiri GI, Enobakhare DA and Thottappilly G. 1986. The importance of colonizing and non-colonizing aphid vectors in the spread of cowpea aphid-borne mosaic virus in cowpea. *Crop Protection*, 5 (6):406-410.

Baidoo PK, Baidoo-Ansah D and. Agbonu I. 2012. Effects of Neem (*Azadirachta indica* A. Juss). Products on *Aphis craccivora* and its predator *Harmonia axyridis* on cowpea. *American Journal of Experimental Agriculture*, 2:198-206.

Baker TA, Nielsen SS, Shade RE and Singh BB. 1989. Physical and chemical attributes of cowpea lines resistant and susceptible to *Callosobruchus maculatus* (F.) (Coleoptera: Bruchidae). *Journal of Stored Products Research*, 25:1-8.

Bateman R, Carey M, Moore D and Prior C. 1993. The enhanced infectivity of *Metarhizium flavoviride* in oil formulations to desert locusts at low humidities. *Annals of Applied Biology*, 122:145-152.

Bell JV. 1974. Mycoses. In: Cantwell, G.E., Ed., Insect Diseases, Marcel Dekker Inc. New York, 185-236 pp.

Bernhard K, Holloway PJ and Burges HD. 1998. Appendix I: A catalogue of formulation additives: Function, nomenclature, properties and suppliers. Formulation of Microbial Biopesticides: Beneficial Microorganisms, Nematodes and Seed Treatments. HD Burges, (ed). Kluwer Academic Publishers, Dordrecht, the Netherlands, pp. 333-365.

Blackman RL and Eastop VF. 2000. Aphids on the world's crops: an identification and information guide (No. Ed. 2). John Wiley & Sons Ltd. 466pp.

Blackman RL and Eastop VF. 2006. Aphids on the world's herbaceous plants and shrubs. Chichester, UK: John Wiley & Sons. 1460 pp.

Blackman RL and Eastop VF. 2007. Taxonomic Issues. In: Van Emdenm HF, Harrington R (eds). Aphids as crop pests. CAB International, Wallingford, 1–29.

Boivin G, Hance T and Brodeur J. 2012. Aphid parasitoids in biological control. *Canadian Journal of Plant Science*, 92 (1):1-12.

Boucias DG, Farmerie WG and Pendland JC. 1998. Cloning and sequencing the cDNA of the insecticidal toxin, Hirsutellin A. *Journal of Invertebrate Pathology*, 72:258-261.

Boxtel JV, Singh BB, Thottappilly G and Maule AJ. 2000. Resistance of cowpea (*Vigna unguiculata* (L.) Walp.) breeding lines to blackeye cowpea mosaic and cowpea aphid-borne mosaic potyvirus isolates under experimental conditions. *Zeitschrift für Pflanzenkrankheiten und Pflanzenschutz*, 107:197-204.

Boucias DG and Pendland JC. 2012. Principles of insect pathology. Springer Science & Business Media, 543 pp.

Brady CM and White JA. 2013. Cowpea aphid (*Aphis craccivora*) associated with different host plants has different facultative endosymbionts. *Ecological Entomology*, 38:433-437.

Braga GU, Flint SD, Miller CD, Anderson AJ, Roberts DW. 2001. Both solar UVA and UVB radiation impair conidial culturability and delay germination in the entomopathogenic fungus *Metarhizium anisopliae*. *Photochemistry and Photobiology*, 74 (5):734-9.

Brooks AJ, De Muro MA, Burree E, Moore D, Taylor M and Wall R. 2004. Growth and pathogenicity of isolates of the fungus *Metarhizium anisopliae* against the parasitic mite, *Psoroptes ovis*: effects of temperature and formulation. *Pest Management Science*, 60:1043-1049.

Burges HD. 1998. Formulation of mycoinsecticides. In: Burges HD, (ed). Formulation of Microbial Biopesticides. Dordrecht: Kluwer Academic Publishers, pp. 131-85.

Butt TM, Harris SG and Powell KA. 1999. Microbial Biopesticides. The European Scene, in Biopesticides Use and Delivery. Hall F R and Menn J J.(eds)Humana Press, Totowa, New Jersey29-44, pp.

Butt TM and Goettel MS. 2000. Bioassays of entomogenous fungi. Bioassays of entomopathogenic microbes and nematodes. In: A. Navon and K.R.S. Ascher (eds). Bioassays of Entomopathogenic Microbes and Nematodes. CABI Publishing CABI CAB International Wallingford UK, 141-195 pp.

Butt TM, Jackson C and Magan N. 2001. Fungi as biocontrol agents: progress, problems and potential. CABI International, Wallingford, 390 pp.

Chandler D, Bailey AS, Tatchell GM, Davidson G, Greaves J and Grant WP. 2011. The development, regulation and use of biopesticides for integrated pest management. Philosophical transactions of the royal society of London. Series B, *Biological Science*, 366 (1573):1987: 98.

Chandra Teja KNP and Rahman SJ. 2016. Characterization and evaluation of *Metarhizium anisopliae* (Metsch.) Sorokin strains for their temperature tolerance *Mycology*, 1-9.

Chaudhary HC and Singh R. 2012. Records of the predators of aphids (homoptera: aphididae) in Eastern Uttar Pradesh. *Journal of Aphidology*, 13.

Chiulele RM, Mwangi G, Tongoona P, Ehlers JD and Ndeve AD. 2011. Assessment of farmers' perceptions and preferences of cowpea in Mozambique. *African Crop Science Conference Proceedings*, 10:311-318.

Citadin CT, Ibrahim AB and Aragão FJ. 2011. Genetic engineering in cowpea (*Vigna unguiculata*): history, status and prospects. *GM Crops*, 2 (3):144-149.

Cohen E, Joseph T and Wassermann-Golan M. 2001. Photostabilization of Biocontrol Agents by Berberine. *International Journal of Pest Management*, 47-67.

Copping LG and Menn JJ. 2000. Biopesticides: a review of their action, applications and efficacy. *Pest Management Science*, 56 (8):651-676.

Cosentino-Gomes D, Rocco-Machado N, Santi L, Broetto L, Vainstein MH, Meyer-Fernandes JR, Schrank A and Beys-da-Silva WO. 2013. Inhibition of ecto-phosphatase activity in conidia reduces adhesion and virulence of *Metarhizium anisopliae* on the host insect *Dysdercus peruvianus*. *Current Microbiology*, 66 (5):467-474.

Costa LB, Rangel DEN, Morandi MAB and Bettiol W. 2012. Impact of UV-B Radiation on *Clonostachys rosea* germination and growth. *World Journal of Microbiology and Biotechnology*, 28: 2497-2504

Damiri BV, Al-Shahwan IM, Al-Saleh MA, Abdalla OA and Amer MA. 2013. Identification and characterization of cowpea aphid-borne mosaic virus isolates in Saudi Arabia. *Journal of Plant Pathology*, 79-85.

Ddamulira G, Santos CAF, Obuo P, Alanyo M and Lwanga CK. 2015. Grain yield and protein content of Brazilian cowpea genotypes under diverse Ugandan environments. *American Journal of Plant Sciences*, 6:2074.

Dassonville N, Thielemans T, Herbener M and Rosemeyer V. 2012. The use of a mix of parasitoids to control all aphid species on protected vegetable crops. IOBC/wprs Bulletin, 88:261-266.

Dent D. 1991. Insect pest management. Wallingford, UK: CAB International, 604 pp.

Dewhirst SY, Pickett JA and Hardie J. 2010. Aphid pheromones. In Vitamins & Hormones Academic Press (Vol. 83, 551-574 pp).

Davies DW, Oelke EA, Oplinger ES, Doll JD, Hanson CV and Putnam DH. 1991. Cowpea. Alternative Field Crops Manual. University of Wisconsin Cooperative for Extension Service, Madison, WI.

Dimbi S, Maniania NK, Lux SA and Mueke JK. 2004. Effect of constant temperatures on germination, radial growth and virulence of *Metarhizium anisopliae* to three species of African tephritid fruit flies. *Biocontrol*, 49:83-94.

Dwivedi SL, Sahrawat K, Upadhyaya H. Mengoni A, Galardini M, Bazzicalupo M, Biondi EG, Hungria M, Kaschuk G, Blair MW and Ortiz R. 2015. Advances in host plant and rhizobium genomics to enhance symbiotic nitrogen fixation in legumes, In: D.L. Sparks (Ed.), Advances in Agronomy, Elsevier Inc, Amsterdam, 1–116 pp.

Egho EO. 2010. Comparative studies on insect species of cowpea (*Vigna unguiculata*) (L) Walp) in two agro-ecological zones during the early cropping season in Delta State, Southern Nigeria. *Agriculture and Biology Journal of North America*, 1 (5):946-949.

Egho EO and Enujeke EC. 2012. Minimizing insecticide application in the control of insect pests of cowpea (*Vigna unguiculata* (L) WALP) in Delta State, Nigeria. *Sustainable Agriculture Research*, 1 (1):87.

Ekesi S, Akpa AD, Onu I, Ogunlan MO. 2000. Entomopathogenicity of *Beauveria bassiana* and *Metarhizium anisopliae* to the cowpea aphid, *Aphis craccivora*. *Archives of Phytopathology and Plant Protection*, 33:171–180.

Espadaler X, Pérez Hidalgo N and Villalobos Muller W. 2012. Ant-aphid relations in Costa Rica, Central America (Hymenoptera: Formicidae; Hemiptera: Aphididae). *Sociobiology*, 59 (3):959.

Eyheraguibel B, Richard C, Ledoigt G and Ter Halle A. 2010. Photoprotection by plant extracts: a new ecological means to reduce pesticide photodegradation. *Journal of Agricultural and Food Chemistry*, 58 (17): 9692-9696.

Ezueh, MI. 1991. Prospects for cultural and biological control of cowpea pests. *International Journal of Tropical Insect Science*, 12:585-592.

Fargues J, Goettel MS, Smits N, Ouedraogo A, Vidal C, Lacey LA, Lomer CJ and Rougier M. 1996. Variability in susceptibility to simulated sunlight of conidia among isolates of entomopathogenic hyphomycetes. *Mycopathologia*, 135:171-181.

Faria MR and Wraight SF. 2007. Mycoinsecticides and mycoacaricides: a comprehensive list with worldwide coverage and international classification of formulation types. *Biological Control*, 43 (3):237-256.

Farrell JAK. 1976. Effects on intersowing with beans on the spread of groundnut rosette virus by *Aphis craccivora* Koch (Hemiptera, Aphididae) in Malawi. *Bulletin of Entomological Research*, 66:331-333.

Ferreira ES, Rodrigues AR, Silva-Torres CS and Torres JB. 2013. Life-history costs associated with resistance to lambda-cyhalothrin in the predatory ladybird beetle *Eriopis connexa*. Agricultural and Forest Entomology, 15 (2):168-177.

33

Ferro DN. 1987. Insect pest outbreaks in agroecosystems. In: Barbosa P, Schultz JC (eds). Insect Outbreaks. San Diego and London: Academic Press. 195-215pp.

Ferron P. 1978. Biological control of insect pests by entomogenous fungi. *Annual Review of Entomology*, 23:409–422.

Ferron P, Fargues J and Riba G. 1991. Fungi as microbial insecticides against pests. Handbook of applied mycology, 2, 665-706 pp.

Flatt T and Weisser WW. 2000. The effects of mutualistic ants on aphid life history traits. *Ecology*, 81 (12):3522-3529.

Food and Agriculture Organization (FAO) 2016.

Gelernter WD, Lomer CJ. 2000. Success in biological control of above-ground insects by pathogens. In: Gurr G, Wratten S (eds). Biological control: measures of success. Kluwer Academic, Dordrecht, The Netherlands, 297–322 pp.

Ghaly AE and Alkoaik FN. 2010. Extraction of protein from common plant leaves for use as human food. *American Journal of Applied Sciences*, 7 (3):331.

Glare TR and Milner RJ. 1991. Ecology of entomopathogenic fungi. In: Arora DK, Ajello L, Mukerji KG (eds) Handbook of applied mycology, vol 2: humans, animals and insects. Marcel Dekker, New York, 547-612 pp.

Glare T, Caradus J, Gelernter W, Jackson T, Keyhani N, Köhl J, Marrone P, Morin L and Stewart A. 2012. Have biopesticides come of age? *Trends in Biotechnology*, 30 (5):250-258.

Goettel MS and Hajek AE. 2000. Evaluation of non-target effects of pathogens used for management of arthropods. In: Wajnberg E, Scott JK, Quimby PC (eds)Evaluating indirect ecological effects of biological control. CABI Publishing, Wallingford, 81–97 pp.

Goławska S, Krzyżanowski R and Łukasik I. 2010. Relationship between aphid infestation and chlorophyll content in Fabaceae species. *Acta Biologica Cracoviensia Series Botanica*, 52 (2):76-80.

Gomez KS, Oosterhuis DM, Rajguru SN and Johnson DR. 2004. Molecular biology and physiology. Foliar antioxidant enzyme responses in cotton after aphid herbivory. *Journal of Cotton Science*, 8:99–104.

Gowtham V, Dilipsundar N, Balaji K and Karthikeyan S. 2016. Study on the effectiveness of pesticides against cowpea aphid (*Aphis craccivora*) Koch. *International Journal of Plant Protection*, 9(1):146-149.

Gullan, PJ and Cranston PS. 1994. The Insects: An Outline of Entomology. Chapman and Hall, London, 491 pp.

Guillon ML. 2003. Regulation of biological control agents in Europe. Roettger U, Reinhold M (eds). In: International symposium on biopesticides for developing countries. CATIE, Turrialba, 143–147 pp.

Guinossi HDM, Moscardi F, Oliveira MCND and Sosa-Gómez DR. 2012. Spatial dispersal of *Metarhizium anisopliae* and *Beauveria bassiana* in soybean fields. *Tropical Plant Pathology*, 37 (1):44-49.

Hajek AE, St. Leger RJ. 1994. Interactions between fungal pathogens and insect hosts. *Annual Review of Entomology* 39 (1):293-322.

Hall AE 2012 Phenotyping cowpeas for adaptation to drought. *Frontiers in Physiology*, 3, 1-8.

Hassan S. 2013. Effect of variety and intercropping on two major cowpea (*Vigna unguiculata* Walp) field pests in Mubi, Adamawa state, Nigeria. *International Journal of Agricultural Research and Development*, 1:108-109.

https://www.marketsandmarkets.com. Markets and markets ltd. Accessed October 2018.

Humber RA. 1991. Fungal pathogens of aphids. In: Peters DC, Webster JA, Chlouber CS (eds). Aphid-plant interactions: Populations to molecules, 45–56. Agricultural Experiment Station, Division of Agriculture, Oklahoma State University, Stillwater

Hunt VL, Lock GD, Pickering SG and Charnley AK. 2011. Application of infrared thermography to the study of behavioral fever in the desert locust. *Journal of Thermal Biology*, 36 (7):443-451.

Hunt VL, Zhong W, McClure CD, Mlynski DT, Duxbury EM, Keith Charnley A and Priest NK. 2016. Cold-seeking behaviour mitigates reproductive losses from fungal infection in Drosophila. *Journal of Animal Ecology*, 85 (1):178-186.

Huynh BL, Ehlers JD, Close TJ, Cisse´ N, Drabo I, Boukar O, Lucas MR, Wanamaker S, Pottorff M and Roberts PA. 2013. Enabling tools for modern breeding of cowpea for biotic stress resistance Translational genomics for crop breeding. Wiley, London, 183–199 pp.

Huynh BL, Ehlers JD, Ndeve A, Wanamaker S, Lucas MR, Close TJ and Roberts P A. 2015. Genetic mapping and legume synteny of aphid resistance in African cowpea (*Vigna unguiculata* L. Walp.) grown in California. *Molecular Breeding*, 35 (1), 36.

Inglis GD, Johnson DL and Goettel MS. 1996. Effects of temperature and thermoregulation on mycosis by *Beauveria bassiana* in grasshoppers. *Biological Control* 7:131–39.

Inglis GD, Johnson DL, Cheng KJ and Goettel MS. 1997. Use of pathogen combinations to overcome the constraints of temperature on entomopathogenic hyphomycetes against grasshoppers. *Biological Control*, 8 (2):143-152.

Inglis GD, Ivie TJ, Duke GM and Goettel MS. 2000. Influence of rain and conidial formulations on persistence of *Beauveria bassiana* on potato leaves and Colorado potato beetle larvae. *Biological Control*, 18:55-64.

Inglis GD, Goettel MS, Butt TM and Strasser H. 2001. Use of hyphomycetous fungi for managing insect pests. In Butt T, Jackson C and Morgan N. (Eds), Fungi as biocontrol agents. Wallingford, UK: CABI, 23-69 pp.

Inglis DG, Jaronski ST, Wraight SP, Beattie A and Watson D. 2002. Use of spray oils with entomopathogens. Spray Oils Beyond 2000: *Sustainable Pest and Disease Management*, 302-312.

Jackai LEN. 1985. Cowpea Entomology Research at IITA and its impact on food production in the tropics. *Nigeria Journal of Entomology*, 6:87-97.

Jackai LEN and Singh SR. 1988. Screening techniques for host plant resistance to insect pests of cowpea. *Tropical Grain Legume Bulletin* 35:2-18.

James RR, Croft BA, Shaffer BT and Lighthart B 1998. Impact of temperature and humidity on host-pathogen interactions between *Beauveria bassiana* and a coccinellid. *Environmental Entomology*, 27: 1506-1513.

Jandricic SE, Filotas M, Sanderson JP and Wraight SP. 2014. Pathogenicity of conidia-based preparations of entomopathogenic fungi against the greenhouse pest aphids *Myzus persicae*, *Aphis gossypii*, and *Aulacorthum solani* (Hemiptera: Aphididae). *Journal of Invertebrate Pathology*, 118:34-46.

Jaronski ST 1997 New paradigms in formulating mycoinsecticides. In: Goss GR, Hopkinson MJ, Collins HM, (eds). Pesticide Formulations and Application Systems. Philadelphia: American Society for Testing and Materials, 99-112 Pp.

Jaronski ST. 2009. Ecological factors in the inundative use of fungal entomopathogens. *BioControl*, 55 (1):159-185.

Jaronski S. 2013. Mass production of entomopathogenic fungi-state of the art. In: mass production of beneficial organisms: invertebrates and entomopathogens. Juan A. Morales-Ramos, M. Guadalupe Rojas and David I. Shapiro-Ilan (eds) Academy, San Diego, California357-413 pp.

Kambhampati S, Vo"lkl W and Mackauer M. 2000. Phylogenetic relationships among genera of Aphidiinae (Hymenoptera: Braconidae) based on DNA sequence of the mitochondrial 16 16S rRNA gene. *Systematic Entomology* 25, 437-445.

Kamphuis LG, Gao L and Singh KB. 2012. Identification and characterization of resistance to cowpea aphid (*Aphis craccivora* Koch) in *Medicago truncatula*. BMC *Plant Biology*, 12:101.

Karungi J, Adipala E, Kyamanywa S, Ogenga-Latigo MW, Oyobo N and Jackai LEN. 2000. Pest management in cowpea. Part 2. Integrating planting time, plant density and insecticide application for management of cowpea field insect pests in Eastern Uganda. *Crop Protection* 19 (4):237-245.

Kaya HK and Vega FE. 2012. Scope and basic principles of insect pathology. Insect Pathology. Academic Press, London, United Kingdom, 1-12 pp.

Kim JS, Roh JY, Choi JY, Shin SC, Jeon MJ and Je YH. 2008. Identification of an entomopathogenic fungus, *Beauveria bassiana* SFB-205 toxic to the green peach aphid, *Myzus persicae*. *International Journal of Industrial Entomology*, 17:211-215.

Lacey LA, Frutos R, Kaya HK and Vail P. 2001. Insect pathogens as biological control agents: Do they have a future? *Biological Control* 21:230-248.

Lattanzio V. Arpaia S, Cardinali A, Di Venere D and Linsalata V. 2000. Role of endogenous flavonoids in resistance mechanism of *Vigna* to aphids. *Journal of Agricultural and Food Chemistry*, 48 (11):5316-5320.

Leng P, Zhang Z, Guangtang P, Zhao M. 2014. Applications and development trends in biopesticides. *African Journal of Biotechnology*, 10:19864–19873.

Li Z, Alves SB, Roberts DW, Fan M, Delalibera I, Tang J, Lopes RB, Faria M and Rangel DEM. 2010. Biological control of insects in Brazil and China: history, current programs and reasons for their success using entomopathogenic fungi. *Biocontrol Science and Technology*, 20: 117–136.

Liu Z, Dai Y, Huang G, Gu Y, Ni J, Wei H and Yuan S. 2011. Soil microbial degradation of neonicotinoid insecticides imidacloprid, acetamiprid, thiacloprid and imidaclothiz and its effect on the persistence of bioefficacy against horsebean aphid *Aphis craccivora* Koch after soil application. *Pest Management Science*, 67(10):1245-1252.

Lopes RB, Pauli G, Mascarin GM and Faria M. 2011. Protection of entomopathogenic conidia against chemical fungicides afforded by an oil-based formulation. *Biocontrol Science and Technology*, 21:125-137.

Maienfisch P, Angst M, Brandl F, Fischer W, Hofer D, Kayser H, Kobel W, Rindlisbacher A, Senn R, Steinemann A and Widmer H. 2001. Chemistry and biology of thiamethoxam: a second generation neonicotinoid. *Pest Management Science*, 57(10):906-913.

Maniania NK. 1993. Evaluation of three formulations of *Beauveria bassiana* (Bals.) Vuill. for control of the stem borer *Chilo partellus* (Swinhoe) (Lep., Pyralidae). *Journal of Applied Entomology*, 115:266–272.

Marrone PG. 2007. Barriers to adoption of biological control agents and biological pesticides. CAB Reviews: perspectives in agriculture, veterinary science, nutrition and natural resources, vol. 2, no. 15. Wallingford, UK: CABI Publishing.

Mayeux A, 1984. The groundnut aphid Biology and control. *Oleagineux*, 39 (8/9):425-434.

Mazid S, Kalida JC and Rajkhowa RC. 2011. A review on the use of biopesticides in insect pest management. *International Journal of Science and Advanced Technology* 1(7):169-178.

Messina FJ, Renwick JAA and Barmcre JL. 1985. Resistance to *Aphis craccivora* (Homoptera: Aphididae) in selected varieties of cowpea. *Journal of Entomological Science*, 20 (2):263-269.

Milner RJ. 1997. *Metarhizium flavoviride* (FI985) as a promising mycoinsecticide for Australian acridids. Goettel MS, Johnson DL (eds). In: Microbial Control of Grasshoppers and Locusts. Memoirs of the Entomological Society of Canada 171:287–300.

Mishra J, Tewari S, Singh S and Arora NK. 2015. Biopesticides: Where We Stand? In: Plant Microbes Symbiosis: Springer India. *Applied Facets*, 37-75 pp.

Mortimore MJ, Singh BB, Harris F and Blade SF. 1997. Cowpea in traditional cropping systems: In B.E. Singh, D.R. Mohan Raj, K.E Dashiel and LEN Jackai (eds), Advances in cowpea research, 99-113 pp.

Mucheru-Muna M, Pypers P, Mugerdi D, Kung'u J, Mugwe J, Merckx R and Vanlauwe B. 2010. A staggered maize–legume intercrop arrangement robustly increases crop yields and economic returns in the highlands of Central Kenya. *Field Crops Research*, 115 (2):132-139.

Murdock LL. 1992. Improving insect resistance in cowpea through biotechnology: Initiatives at Purdue University, USA. In: Thottappilly G, Monti B, Mohan Raj DR and Moore AW (eds) Biotechnology: Enhancing Research on Tropical Crops in Africa, 313-320 pp. CTA-IITA Publication.

Mweke A, Ulrichs C, Maniania KN, Ekesi S. 2016. Integration of entomopathogenic fungi as biopesticide for the management of cowpea aphid (*Aphis craccivora* Koch). *African Journal of Horticultural Science*, 9:14–31.

Nabirye J, Nampala P, Ogenga-Latigo MW, Kyamanywa S, Wilson H, Odeke V, Iceduna C and Adipala E. 2003. Farmer-participatory evaluation of cowpea integrated pest management (IPM) technologies in Eastern Uganda. *Crop Protection*, 22 (1):31-38.

Nampala P, Kyamanywa S, Ogenga-Latigo MW, Adipala E, Karungi J, Oyobo N and Jackai, LEN. 1999. Integrated management of major field pests of cowpea in eastern Uganda. *African Crop Science Journal*, 7:79-486.

Nampala P, Ogenga-Latigo MW, Kyamanywa S, Adipala E, Oyobo N and Jackai LEN. 2002. Potential impact of intercropping on major cowpea field pests in Uganda. *African Crop Science Journal*, 10 (4):335-344.

Nielsen SS, Brandt WE and Singh BB. 1993. Genetic variability for nutritional composition and cooking time of improved cowpea lines. *Crop Science*, 33:469-472.

Nussenbaum AL, Lewylle MA, Lecuona RE. 2013. Germination, radial growth and virulence to boll weevil of entomopathogenic fungi at different temperatures. *World Applied Sciences Journal*, 25:1134–1140.

Ng NQ 1995 Cowpea *Vigna unguiculata* (Leguminosae-Papilionoideae). In: Smart J, Simmonds N.W. (eds) Evolution of crop plants 2nd edition. Longman, Harlow UK, 326-332 pp.

Ofuya TI. 1988a. Resistance of some cowpea varieties to the cowpea aphid, *Aphis craccivora* Koch (Homoptera: Aphididae) under field and screenhouse conditions in Nigeria. *Tropical Pest Management*, 34:445-447.

Ofuya TI. 1988b. Antibiosis in some cowpea varieties resistant to the cowpea aphid, *Aphis craccivora* Koch (Homoptera: Aphididae). *International Pest Control*, 30:68-69.

Ofuya TI and Akingbohungbe AE. 1988. Functional and numerical responses of *Cheilomenes lunata* (Fabricius) (Coleoptera: Coccinellidae) feeding on the cowpea aphid, *Aphis craccivora* Koch (Homoptera: Aphididae). *International Journal of Tropical Insect Science*, 9, 543-546.

Ofuya TI. 1989. Studies on infestation, occurrence, growth and survival of *Aphis craccivora* Koch on cowpea and other alternative hosts in Nigeria. Nigerian. *Journal of Basic and Applied Sciences*, 3:19-25.

Ofuya TI. 1991. Aspects of the ecology of predation in two coccinellid species on the cowpea aphid in Nigeria. Behaviour and Impact of Aphidophaga. SPB Academic Publishing, The Hague, 213-220 pp.

Ofuya TI. 1995. Studies on the capability of *Cheilomenes lunata* (Fabricius) (Coleoptera: Coccinellidae) to prey on the cowpea aphid, *Aphis craccivora* Koch Homoptera: Aphididae) in Nigeria. *Agriculture Ecosystems and Environment*, 52 (1):35-38.

Ofuya TI. 1997. Control of the bean aphid *Aphis craccivora* Koch (Homoptera: Aphididae) in cowpea, *Vigna unguiculata* (L.) Walp. *Integrated Pest Management Reviews*, 2:199-207.

Oiye SO, Shiundu KM and Oniang'o RK. 2009. The Contribution of African leafy vegetables to vitamin A intake and the influence of income in rural Kenya. *African Journal of Food and Nutritional Development*, 9:1- 6.

Olson S. 2015. An analysis of the biopesticide market now and where it is going. Outlooks *on Pest Management*, 26 (5):203-206.

Ortiz-Urquiza A and Keyhani NO. 2013. Action on the surface: entomopathogenic fungi versus the insect cuticle. *Insects*, 4:357-374.

Owolabi AO, Ndidi US, James BD and Amune FA. 2012. Proximate, antinutrient and mineral composition of five varieties (improved and local) of cowpea, *Vigna unguiculata*, commonly consumed in Samaru community, Zaria-Nigeria. *Asian Journal of Food Science and Technology*, 4(2):70-72.

Pell JK, Eilenberg J, Hajek AE and Steinkraus DC. 2001. Biology, ecology and pest management potential of Entomophthorales. In: Butt TM, Jackson C, Magan N (eds) Fungi as biocontrol agents: progress, problems and potential. CAB International, Wallingford, 71–153 pp.

Pickett JA, Butt TM, Doughty KJ, Wallsgrove RM, Williams IH. 1995. Minimizing pesticide input in oilseed rape by exploiting natural regulatory processes. Plenary lecture. In: Proceedings of the GCIRC 9th International Rapeseed Congress, Cambridge, UK, 4–7 July.

Powell W and Pickett JA. 2003. Manipulation of parasitoids for aphid pest management: progress and prospects. *Pest management Science*, 59 (2):149-155.

Pupin AM, Messias CL, Piedrabuena AE and Roberts DW. 2000. Total lipids and fatty acids of strains of *Metarhizium anisopliae*. *Brazilian Journal of Microbiology*, 31:121–128.

Radha R. 2013. Comparative studies on the effectiveness of pesticides for aphid control in cowpea. *International Journal of Plant Protection*, 9 (1):1-7

Ramoska WA. 1984. The influence of relative humidity on *Beauveria bassiana* infectivity and replication in the chinch bug, *Blissus leucopterus*. *Journal of Invertebrate Pathology*, 43 (3):389-394.

Rangel DE, Anderson AJ and Roberts DW. 2008. Evaluating physical and nutritional stress during mycelial growth as inducers of tolerance to heat and UV-B radiation in *Metarhizium anisopliae* conidia. *Mycological Research*, 112:1362-1372.

Rangel DE, Fernandes ÉK, Dettenmaier SJ and Roberts DW. 2010. Thermotolerance of germlings and mycelium of the insect-pathogenic fungus *Metarhizium* spp. and mycelial recovery after heat stress. *Journal of Basic Microbiology*, 50 (4):344-350.

Reddy NP, Khan A, Akbar P, Devi KU, Victor JS and Sharma HC. 2008. Assessment of the suitability of tinopal as an enhancing adjuvant in formulations of the insect pathogenic fungus *Beauveria bassiana* (Bals.) Vuillemin. *Pest Management Science*, 64:909-915.

Rizk AM. 2011. Effect of strip-management on the population of the aphid, *Aphis craccivora* Koch and its associated predators by intercropping faba bean, *Vicia faba* L. with coriander, *Coriandrum sativum* L. *Egyptian Journal of Biological Pest Control*, 21 (1):81-87.

Roberts DW and St Leger RJ. 2004. *Metarhizium* spp., cosmopolitan insect-pathogenic fungi: mycological aspects. *Advances in Applied Microbiology*, 54:1-70.

Roddam LF and Rath AC. 1997. Isolation and characterization of *Metarhizium anisopliae* and *Beauveria bassiana* from subantartic Macquarie Island. *Journal of Invertebrate Pathology*, 69:285–288.

Rodrigues IMW, Forim MR, da Silva MFGF, Fernandes JB and Filho AB. 2016. Effect of ultraviolet radiation on fungi *Beauveria bassiana* and *Metarhizium anisopliae*, pure and encapsulated, and bio-insecticide action on *Diatraea saccharalis*. *Advances in Entomology*, 4, 151-162.

Roy HE, Steinkraus DC, Eilenberg J, Hajek AE and Pell JK. 2006. Bizarre interactions and endgames: entomopathogenic fungi and their arthropod hosts. *Annual Review of Entomology*, 51:331-357.

Roy HE, Brodie EL, Chandler D, Goettel MS, Pell JK, Wajnberg E and Vega F. 2010. Deep space and hidden depths: understanding the evolution and ecology of fungal entomopathogens. *Biological* control, 55:1–6.

Roy HE, Brodie EL, Chandler D, Goettel MS, Pell JK, Wajnberg E and Vega F. 2010. Deep space and hidden depths: understanding the evolution and ecology of fungal entomopathogens. *Biological Control*, 55:1–6.

Saidi M, Itulya FM, Aguyoh JN and Mshenga PM. 2010. Yields and profitability of a dual-purpose sole cowpea and cowpea-maize intercrop as influenced by cowpea leaf harvesting frequency. *Journal of Agricultural and Biological Science*, 5:65-71.

Samson RA, Evans HC and Latg JP. 1988. Atlas of entomopathogenic fungi.''Springer Science and Business Media. Berlin, Heidelberg, New York.

Samuels RL, Reynolds SE and Charnley AK. 1988. Calcium-channel activation of insect muscle by destruxins, insecticidal compounds produced by the entomopathogenic fungus *Metarhizium anisopliae*. Comparative Biochemistry and Physiology Part C: *Comparative Pharmacology* 90:403–412.

Santos MP, Dias LP, Ferreira PC, Pasin LA and Rangel DE. 2011. Cold activity and tolerance of the entomopathogenic fungus *Tolypocladium* spp. to UV-B irradiation and heat. *Journal of Invertebrate Pathology*, 108 (3):209-213.

Schipanski ME and Drinkwater LE. 2012. Nitrogen fixation in annual and perennial legume-grass mixtures across a fertility gradient. *Plant and Soil*, 357(1-2):147-159.

Selvaraj K, Kaushik HD, Gulati R and Sharma SS. 2010. Bioefficacy of *Beauveria bassiana* (Balsamo) Vuillemin against *Hyadaphis coriandri* (Das) on coriander and *Aphis craccivora* Koch on fenugreek. *Journal of Biological Control*, 24 (2):142-146.

Shah PA and Pell JK. 2003. Entomopathogenic fungi as biological control agents. *Applied Microbiology and Biotechnology*, 61 (5):413-423.

Shahid AA, Rao QA, Bakhsh A and Husnain T. 2012. Entomopathogenic fungi as biological controllers: new insights into their virulence and pathogenicity. *Archives of Biological Sciences*, 64 (1):21-42.

Smith CM and Clement SL. 2012. Molecular bases of plant resistance to arthropods. *Annual Review of Entomology*, 57:309-328.

Smith CM and Chuang WP. 2014. Plant resistance to aphid feeding: behavioral, physiological, genetic and molecular cues regulate aphid host selection and feeding. *Pest Management Science*, 70:528–540.

Smith PT and Kambhampati S. 2000 Evolutionary transitions in Aphidiinae (Hymenoptera: Braconidae). In: Austin A, Dowton M. (Eds.), Hymenoptera: Evolution, Biodiversity and Biological Control. CSIRO Publishing, Australia, 104–111 pp.

Sorensen JT. 2009. Aphids. In Encyclopedia of Insects. Resh, V.H. and Cardé, R.T. (eds) (Second Edition) Academic Press, 27-31 pp.

Stary P. 1970. Biology of aphid parasites with respect to integrated control. Series ' Entomologica, 6. Junk Publishers, Hague, The Netherlands.

Stary' P .1988. Aphidiidae. In: A K Minks and P Harrewijn, (eds). Aphids: their biology, natural enemies and control, Volume 2B of World Crop Pests. Elsevier, New York, NY, 171-184 pp.

Stary' P. 2013. Aphid parasites (Hymenoptera, Aphidiidae) of the central Asian area. Springer Science and Business Media.

Steinhaus EA. 1956. Microbial control-the emergence of an idea. A brief history of insect pathology through the nineteenth century. *Hilgardia* 26:107-160.

Thakore Y 2006 The biopesticide market for global agricultural use. *Industrial Biotechnology*, 2 (3):194-208.

Thevelein JM. 1984. Regulation of trehalose mobilization in fungi. *Microbiological Reviews*, 48 (1), 42.

Thomas MB and Jenkins NE. 1997. Effects of temperature on growth of *Metarhizium flavoviride* and virulence to the variegated grasshopper, *Zonocerus variegatus*. *Mycological Research*, 101:1469–1474.

Trehan I, Benzoni NS, Wang AZ, Bollinger LB, Ngoma TN, Chimimba UK, Stephenson KB, Agapova SE, Maleta KM and Manary MJ. 2015. Common beans and cowpeas as complementary foods to reduce environmental enteric dysfunction and stunting in Malawian children: study protocol for two randomized controlled trials. *Trials*, 16 (1):520.

Trenbath BR. 1993. Intercropping for the management of pests and diseases. *Field Crops Research*, 34(3-4):381-405.

Van Emden HF. 1991. The role of host plant resistance in insect pest mis-management. *Bulletin of Entomological Research*, 81:123-126.

Vega FE and Blackwell M (eds). 2005. Insect-Fungal Associations: Ecology and Evolution. New York: Oxford University Press, Inc, 332 pp.

Vega FE, Meyling NV, Luangsa-ard JJ and Blackwell M. 2012. Fungal Entomopathogens. In: Vega FE and Kaya HK (eds.), Insect pathology (2nd ed.), Amsterdam, the Netherlands: Elsevier, 171-220 pp.

Vidal C, Fargues J and Lacey LA. 1997. Intraspecific variability of *Paecilomyces fumosoroseus:* Effect of temperature on vegetative growth. *Journal of Invertebrate Pathology*, 70:18-26.

Vollhardt IM, Tscharntke T, Wäckers FL, Bianchi FJ and Thies C. 2008. Diversity of cereal aphid parasitoids in simple and complex landscapes. *Agriculture, Ecosystems and Environment*, 126 (3-4):289-292.

Völkl W, Mackauer M, Pell J and Brodeur J. 2007. Predators, parasitoids and pathogens. In: Van Emden HF and Harrington R. (Ed.) Aphids as crop pests, pp. Wallingford, UK CABI187-233 pp.

Vu VH Hong S and Kim K. 2007. Selection of entomopathogenic fungi for aphid control. *Journal of Bioscience and Bioengineering*, 104:498-505.

Wabule MN, Ngaruiya PN, Kimmins FK and Silverside PJ. 2004. (eds). Registration for biocontrol agents in Kenya. Proceedings of the PCPB/KARI/DFID CPP workshop, 14th-16th May, 2003, Nakuru, Kenya.

Waddington SR, Li X, Dixon J, Hyman G and De Vicente MC. 2010. Getting the focus right: production constraints for six major food crops in Asian and African farming systems. *Food Security*, 2 (1):27-48.

Wei JN, Bai BB, Yin TS, Wang Y, Yang Y, Zhao LH, Kuang RP and Xiang RJ. 2005. Development and use of parasitoids (Hymenoptera: Aphidiidae and Aphelinidae) for biological control of aphids in China. *Biocontrol Science and Technology*, 15:533-551.

Wraight SP and Carruthers RL. 1999. Production, delivery, and use of mycoinsecticides for control of insect pests of field crops. In: Hall FR, Menn JJ, (eds). Methods in Biotechnology, Vol. 5: Biopesticides: Use and Delivery. Totowa, New Jersey, USA: Humana Press, 233-69 pp.

Wraight SP, Jackson MA and Dekock SL. 2001. Production, stabilization and formulation of fungal biocontrol agents. In: Butt TM, Jackson C, Magan N, (eds). Fungal biocontrol agents-Progress, Problems and potential. Wallingford: CAB International, 253-287 pp.

Wraight SP and Ramos ME. 2002. Application parameters affecting field efficacy of *Beauveria bassiana* foliar treatments against Colorado potato beetle *Leptinotarsa decemlineata*. *Biological Control*, 23:164-78.

Xiao G, Ying SH, Zheng P Wang ZL, Zhang S, Xie XQ, Shang Y, Leger RJS, Zhao GP, Wang C and Feng MG. 2012. Genomic perspectives on the evolution of fungal entomopathogenicity in *Beauveria bassiana*. *Scientific Reports* 2, 483.

Yeo H, Pell JK, Alderson PG, Clark SJ and Pye BJ. 2003. Laboratory evaluation of temperature effects on the germination and growth of entomopathogenic fungi and on their pathogenicity to two aphid species. *Pest Management Science*, 59:156-165.

Zimmermann G. 2007 Review on safety of the entomopathogenic fungus *Metarhizium anisopliae*. *Biocontrol Science and Technology*, 17 (9):879-920.

3.0 EVALUATION OF THE ENTOMOPATHOGENIC FUNGI *METARHIZIUM ANISOPLIAE*, *BEAUVERIA BASSIANA* AND *ISARIA* SP. FOR THE MANAGEMENT OF *APHIS CRACCIVORA*

Published as: Evaluation of the entomopathogenic fungi *Metarhizium anisopliae*, *Beauveria bassiana* and *Isaria* sp. for the management of *Aphis craccivora*
Mweke A, Ulrichs C, Nana P, Akutse KS, Fiaboe K.K.M, Maniania N.K. and Ekesi S 2018. *Journal of Economic Entomology*, 111(4), 1587–1594.

3.1 Abstract

Cowpea, *Vigna unguiculata*, is an important indigenous vegetable and grain legume in the tropics where it represents a major diet component. Cowpea aphid, *Aphis craccivora* is a major pest causing up to 100% yield losses. Aiming at establishing alternative approach to synthetic insecticides, we evaluated the pathogenicity of 23 fungal isolates including *Metarhizium anisopliae*, *Beauveria bassiana* and *Isaria* sp. against adult *A. craccivora* in the laboratory. Adult apterous aphids were sprayed with conidial suspensions titred at 1×10^8 conidia ml^{-1} for pathogenicity tests while 1×10^4, 1×10^5, 1×10^6, 1×10^7 and 1×10^8 conidia ml^{-1} were used in dose response bioassays. All the fungal isolates were found pathogenic to *A. craccivora*, causing mortality of between 34.5 and 90%. The lethal 50% mortality time (LT_{50}) values varied between 3.3 and 6.3 days, with the best isolates being ICIPE 62, ICIPE 41 and ICIPE 644. The lethal concentration mortality (LC_{50}) values were 2.3×10^6, 1.3×10^8 and 1.3×10^9 for ICIPE 62, ICIPE 41 and ICIPE 644, respectively. *M. anisopliae* isolate ICIPE 62 produced more conidia on aphid cadavers (4.5×10^7) than ICIPE 41 (2.7×10^7) and ICIPE 644 (2.1×10^7) 6 days post-treatment. Relative potency comparison showed that ICIPE 62 was more potent than the other two isolates. In the screenhouse, conidia of ICIPE 62 significantly reduced *A. craccivora* population compared to control but there was no significant difference between emulsifiable and aqueous formulations. Small-holder leafy vegetable producers could gain more profits using fungal-based biopesticides in Aphid-IPM strategies, leading to reduction of pre-harvest intervals after their application compared to synthetic insecticides.

Key words: entomopathogenic fungi, cowpea aphid, pathogenicity, mortality, biological control.

3.2 Introduction

Cowpea, *Vigna unguiculata* ((L.) Walp; Fabales: Fabaceae), is an important food legume and vegetable crop in rural and urban areas in Africa (Saidi et al., 2010). The crop does well in the marginal rainfall areas because it is well adapted to dry climate and suitable for a variety of intercropping systems (Singh et al., 2000; Muchero et al., 2009; Boukar et al., 2011). Cowpea is grown for the seeds, the pods or leaves that are consumed as green vegetables and animal feed as well as for incorporation as green manure for soil fertility management (Singh and Sharma1996; Fatokun 2000; Hall 2004; Baidoo and Mochiah, 2014). Cowpea leaves contain high protein content (29-41%) (Ehlers and Hall, 1997) and meet nutritional needs of a large population in rural areas. Cowpea yields in sub-Saharan Africa are low and range between 100 and 250 kg/ha (Omongo et al,. 1997, Baidoo et al., 2012) while in other parts of the world yields of 3000 kg/ha have been recorded (Rusoke and Rubaihayo, 1994; Hall 2012). A plethora of indigenous and invasive insect pests and diseases as well as abiotic factors contribute to these low yields. Among the insect pests of cowpea, the cowpea aphid, *Aphis craccivora* (Koch; Hemiptera: Aphididae), is one of the most economically important pests causing yield losses of 20-100% (Nampala et al,. 1999; Obopile, 2006). *Aphis craccivora* causes direct and indirect damage on cowpea through sucking of plant sap and transmission of more than 30 plant viruses while the honey dew produced by the aphids on leaves interferes with photosynthetic process of the plants (Blackman and Eastop, 2000). During heavy infestation, the plants become stunted and result in delayed flowering and reduced yields (Blackman and Eastop, 2000). Although integrated pest management (IPM) that consists of the use of resistant varieties, intercropping, good crop husbandry, manipulation of planting dates and minimum application of synthetic chemical insecticides is recommended for the management of this insect pest (Abate and Ampofo 1996, Asante et al,. 2001, Jackai and Asante, 2003) synthetic chemicals control are commonly used. The latter is associated with environmental and health risks, resistance development, as well as detrimental effects on beneficial organisms (Ofuya1997; Baidoo et al., 2012).

Increasing pressure from consumers and retailers of agricultural produce to reduce pesticide residues has accelerated search for alternative pest management options such as IPM approaches and development and application of microbial insecticides (Gwynn and Maniania 2010; Chandler et al., 2011; Leng et al., 2011).

Although microbial insecticides are slow in killing target pest and usually used as an IPM component, they are attractive to horticultural farmers because not only they leave no residues, but also have shorter pre-harvest interval, reduced negative environmental impacts and are safer to users compared to pesticides derived from synthetic molecules (Lacey et al., 2001; Chandler et al., 2008).

Among the microbial insecticides, entomopathogenic fungi (EPF), whose infection of the host is through the cuticle and do not need to be ingested like bacteria, viruses and microsporidia, are most suitable for sap-feeding insects such as aphids (Hajek and St. Leger, 1994; Vu et al., 2007). Aphid species including *A. craccivora* are generally susceptible to EPF (Ekesi et al., 2000; Sahayaraj and Borgio 2010; Saranya et al., 2010; Bayissa et al., 2016a) and several of them have been developed as mycoinsecticides (Faria and Wraight, 2007).

45

The lack of guidance on import and release of biological control agents in most African countries as well as restrictions on the use of biological diversity between countries has fostered research for local virulent isolates of EPF that could be used for biocontrol of insect pests (Cherry and Gwynn, 2007; Gwynn and Maniania, 2010). In addition, consumers' awareness on the risks associated with the use of synthetic pesticides and on the benefits of adopting safe pest control products have strengthened research on development of biopesticides (Chandler et al., 2011; Leng et al., 2011). There are many steps in developing EPF as mycoinsecticides and the most important include selection of virulent strains and or isolates and evaluation of their performance under different environmental conditions as well as selection of cost-effective production technology and effective formulation (Jenkins et al., 1998; Boyetchko et al., 1999; Montesinos, 2003; Jaronski, 2010).

Key EPF species used in agricultural pest management include *Metarhizium anisopliae* ((Metschn.) Sorokin; Hypocreales: Clavicipitaceae), *Beauveria bassiana* ((Bals.) Vuill.; Hypocreales: Cordycipitaceae) and *Isaria* sp (Hypocreales: Cordycipitaceae). The potential of various *M. anisopliae* isolates in *A. craccivora*'s management has been largely documented (Ekesi et al., 2000, Sahayaraj and Borgio 2010, Saranya et al., 2010, Bayissa et al., 2016a). The pathogenicity of *M. anisopliae* isolate ICIPE 62 has been reported on various aphid species such as *Brevicoryne brassicae* (Linnaeus; Hemiptera: Aphididae), *Lipaphis pseudobrassicae* (Davis; Hemiptera: Aphididae) and *Aphis gossypii* (Glover; Hemiptera: Aphididae), (Bayissa et al., 2016a). Although *M. anisopliae* ICIPE 62 has been developed as biopesticide commercially known as Met 62® by the International Centre of Insect Physiology and Ecology (*icipe*) in collaboration with Real IPM Kenya against major vegetable arthropod pests including aphid species (http://www.realipm.com/), its efficacy against *A. craccivora* has never been assessed. This study, therefore, reported for the first time the pathogenicity of ICIPE 62 against *A. craccivora* that could promote the use of this biopesticide in cowpea production systems. The development and use of EPF-based biopesticides is not only compatible with other IPM strategies applied in insect pest management, but also economically viable to small-holder vegetable farmers (Gwynn and Maniania, 2010). In addition, small-holder leafy vegetable producers could gain more profits using fungal-based biopesticides, since this reduces the pre-harvest intervals after application compared to synthetic insecticides. The objectives of the present study were therefore to screen fungal isolates of *M. anisopliae*, *B. bassiana* and *Isaria* sp. for their virulence against the *A. craccivora* in order to select the most virulent isolate(s) that could be further developed as mycoinsecticide and to evaluate the performance of the selected isolate against cowpea aphid in the screenhouse.

3.3 Materials and methods

3.3.1 Insects

The initial *A. craccivora* stock used in the experiment was obtained from *icipe*'s Duduville campus, Kasarani, Nairobi (1.2219°S, 36.8967°E and attitude of 1590 meters above sea level (msl) collected on infested Ex-Luanda cowpea land race. The insects were thereafter reared on the same host plant for five generations prior to the experiments following the rearing method previously described by Ekesi et al., 2000. The insects were reared on five to six weeks old potted cowpea plants (Ex-Luanda land race) in glass cages in the laboratory under 27–28 °C, 70% relative humidity and photoperiod of 12:12 L:D. The laboratory insects' stock was renewed every three months with the aim of maintaining insect vigour. To obtain insects of the same age, female apterous aphids were introduced into clean and uninfested potted plants in cages and allowed to reproduce for 24h then removed. The nymphs were then reared to maturity for five days and used in the experiment.

3.3.2 Fungal isolate cultures

All the fungal isolates used in this experiment were obtained from the *icipe*'s Arthropod Germplasm Centre. Details on original source of the isolates, location and year of isolation are summarized in Table 1. The isolates were cultured on Sabouraud Dextrose Agar (SDA) in 90-mm Petri dishes at 26 ± 2 °C in darkness. *Metarhizhium anisopliae* isolates were harvested after two weeks while *B. bassiana* and *Isaria* sp. were harvested after three weeks. Conidia were harvested by scraping the surface of the sporulated cultures with a sterile spatula. The inoculum was suspended in 10 ml of sterile distilled water with 0.05% Triton X-100 in glass bottle containing glass beads. The bottles were vortexed for five min to produce homogeneous conidial suspension. The spore concentration of each isolate was determined using an improved Neubauer hemocytometer under light microscope. The required conidial suspension with a standard concentration of 1×10^8 conidia ml^{-1} was obtained for the 23 entomopathogenic fungal isolates through serial dilutions. Viability of the isolates was determined by spread-plating 100 µl of conidial suspension titrated at 3×10^6 conidia ml^{-1} on SDA plates. The plates were incubated at 26 ± 2°C in darkness and examined after 18h. The percentage germination was determined by counting approximately 100 germinated conidia under a light microscope (400x). The conidia were scored as viable when the germ tubes were two times the diameter of the propagule (Goettel et al., 2000).

3.3.3 Screening of fungal isolates for virulence against *Aphis craccivora*

A total of 23 isolates (14 isolates of *M. anisopliae sensu lato*, eight of *B. bassiana sensu lato* and one of *Isaria* sp. were evaluated for their pathogenicity against *A. craccivora*. Whole cowpea leaves were sterilized with 1% sodium hypochlorite and rinsed three times in distilled water for approximately three minutes and allowed to dry in sterile laminar flow chamber after which 20 five-day old apterous adult *A. craccivora* were introduced onto sterilized plastic dishes (11.3 × 4 cm) (diam. x depth) and allowed to settle on the leaves before treatment. Four plastic dishes containing the insects and the leaves (considered as a replicate) were sprayed with 10 ml fungal suspension of each isolate titred to 1×10^8 conidia ml^{-1} using Burgerjon's (1956) spray tower. The spray tower rotates at the base to ensure homogenous distribution of conidial suspension on the plates containing the test insects. Control insects were sprayed with sterile distilled water containing 0.05% Triton X-100. The plastic dishes containing test-insects were allowed to dry for five min to remove excess moisture then covered with lids with apertures (300 x 300 μ) to allow circulation of air. All the treatments were maintained in an incubator at 26 ± 2 °C. The fungal isolates were bioassayed in a group of four to five isolates and each isolate replicated four times. Mortality data were collected daily for seven days after inoculation. Dead aphids were surface-sterilized for one minute with 2.5% sodium hypochlorite and 70% alcohol and thereafter rinsed three times with sterile distilled water. The dead sterilized insects were then placed in Petri dishes lined with moist filter paper to favour the growth of EPF on the surface of the cadaver. Petri dishes were sealed with Parafilm and incubated at 26 ± 2 °C. Death due to fungal isolates was confirmed by observing mycosis on dead insects under a dissecting microscope (Leica microscope).

3.3.4 Dose-mortality response bioassays

The isolates with the shorted lethal time mortality values (LT$_{50s}$) were selected for dose-mortality response bioassays. Twenty five-day old apterous adults were sprayed with 10 ml of each isolate with the following five doses (1×10^4; 1×10^5; 1×10^6; 1×10^7 and 1×10^8 conidia ml^{-1}) using Burgerjon's (1956) spray tower as described above. Control insects were sprayed with sterile distilled water containing 0.05% Triton X-100. The experiment was replicated 3 times and each replicate consisted of 20 test insects per treatment. Mortality data were recorded for seven days and mortality due to fungal infection was carried out using the procedure described above.

3.3.5 Assessment of conidia production on bioassayed *Aphis craccivora*

Based on the screening and dose response bioassay results, the three most potent isolates- two of *M. anisopliae* (ICIPE 41 and ICIPE 62) and one of *B. bassiana* (ICIPE 644) which recorded highest mortality within the shortest time (LT$_{50}$) were selected to assess conidia production level on aphid cadavers using the concentration of 10×10^8 conidial ml^{-1}. Conidial production was assessed on day three, six and nine post-treatment. Three mycosed aphids per treatment were picked from each of the three replicates and dried in an oven at 30 ± 2 °C for 30 min. They were then transferred into universal bottles containing 10 ml distilled water with 0.05% Triton X-100 and vortexed for five minutes. Conidial concentrations were determined using a Neubauer counting chamber (Inglis et al., 2012).

3.3.6 Performance of selected fungal isolate under screenhouse conditions

Based on short LT_{50} and low LC_{50} values and a high conidial production on cowpea aphid cadavers, *M. anisopliae* isolate ICIPE 62 was selected for screenhouse evaluation. Conidial suspension at a concentration of 1×10^8 conidia ml^{-1} was formulated in aqueous formulation (Triton X-100 + water at the following ratio of 0.05:99.95) and emulsifiable formulation (Triton X-100 + Corn oil + water at the ratio of 0.05:0.1:99.85). Glycerin (0.1 %), nutrient agar (0.1%) and molasses (0 5%) were added to each formulation as protectants against UV light (Maniania, 1993). Cowpea plants (Ex-Luanda) were planted in (15 cm × 20 cm) plastic pots. Four seeds per pot were directly sown, thinned to two plants per pot after two weeks then transferred into cages (100 × 100 cm) with fine mesh to allow circulation of air but small enough not to allow escape of the aphids. *Aphis craccivora* used in this study were reared in the laboratory as previously described. One hundred adult apterous *A. craccivora* were introduced into each cage containing two plants and maintained in a screenhouse (5 m × 10 m) at temperature ranging between 25 to 29 °C and 80 to 90% RH and 12: 12 photoperiods (Bayissa et al., 2016a). The insects were allowed to multiply and attain stable age distribution for five days (Banken, 1996). Treatments were randomly allocated before application and initial aphid population (N_0) was determined by destructive sampling of one plant from each cage. This was done to determine the aphid population growth rate based on instantaneous rate of increase (r_i) for each treatment (Birch, 1948). A hand sprayer discharging fine droplets of 41μm diameter was used to spray the infested plants in each cage in the evening to reduce effect of sun light. Control treatments were sprayed with either sterile distilled water containing 0.05% Triton X-100 or 0.1% corn oil with the above listed protectants but without conidia.

To assess the efficacy of the treatments, 20 apterous adult *A. craccivora* were randomly collected from treated plants 20 min post-application in each cage using a soft brush and immediately placed on plastic dishes (11.3×4 cm) lined with moist filter paper containing sterilized cowpea leaves serving as food. They were maintained in an incubator at 26 ± 2 °C; 60 ± 5% RH and 12:12 photoperiod. Mortality of the insects were recorded daily for seven days and dead aphids were observed for mycosis under dissecting microscope. Aphid population growth rate was determined by removing all the treated plants from the cages after five days and counting all the aphids. Each treatment was replicated four times and the experiment was conducted once.

3.3.7 Statistical analysis

Percentage germination data were analyzed using binomial regression analysis of the generalized linear model (GLM) and means were separated using Tukey HSD test. Percentage aphid mortality data were first corrected for natural mortality (Abbott, 1925), followed by normality test (Shapiro and Wilk, 1965), then arcsine transformed before being subjected to Analysis of Variance (ANOVA) using R software (R Core, 2013). Means mortality for different fungal species and isolates were compared using Tukey HSD test. Lethal time and lethal concentration values were determined for each replicate using the probit analysis method for correlation. Data on conidial production on aphid cadavers were tested for normality using Shapiro-Wilk (Shapiro and Wilk, 1965) before analysis and were fitted to GLM using negative binomial regression analysis. To estimate aphid population per plant in the screenhouse count, data were fitted into GLM using negative binomial regression analysis. Aphid population growth rate was estimated using instantaneous rate of increase (r_i) from the following equation: $r_i = \ln(N_t/N_0)/t$, where: N_0 represents the initial number of aphids before treatment; N_t represents the aphid population at the end of time t (days); r_i positive values indicates population growth, r_0 values shows no growth or decline in population, while negative r_i values indicates declining population. The means data on aphid population after treatment were separated using Tukey HSD test after being subjected to ANOVA. Two sample t-test procedure was used to compare the emulsifiable and aqueous formulations means. All the Analysis were carried out using R statistical software procedure.

3.4 Results

3.4.1 Pathogenicity of fungal isolates against *Aphis craccivora*

Conidial germination of the different fungal isolates was more than 90 %, except isolate ICIPE 30 that had 82.4% conidial germination (Table 3.1).

Table 3.1: Identity of fungal isolates used in the study and their percent germination on SDA at 24-28 °C

Fungal species	Isolate	Source	Locality/Country	Year of isolation	Germination ± SE* (%)
M. a.	ICIPE 7	Amblyomma variegatum	Rusinga (Kenya)	1996	92.7 ± 1.9efg
	ICIPE 18	Soil	Mbita (Kenya)	1989	93.7 ± 0.6def
	ICIPE 20	Soil	Migori (Kenya)	1989	95.2 ± 0.4cde
	ICIPE 30	Busseola fusca	Kendubay (Kenya)	1989	82.4 ± 4.7h
	ICIPE 40	Soil	Kitui (Kenya)	1990	94.2 ± 1.6def
	ICIPE 41	Soil	Lemba (D.R. Congo)	1990	94.2 ± 2.2ab
	ICIPE 62	Soil	Matete (DRC)	1990	98.2 ± 1.2a
	ICIPE 63	Soil	Matete (DRC)	1990	99.2 ± 0.5a
	ICIPE 68	Soil	Matete (DRC)	1990	98.7 ± 0.5abcd
	ICIPE 69	Soil	Matete (DRC)	1990	97.1 ± 0.9abcd
	ICIPE 74	Soil	Mtwapa (Kenya)	1990	97.5 ± 1.0a
	ICIPE 78	Temnoschoita nigroplagiata	Ugoe (Kenya)	1990	98.3 ± 0.9a
	ICIPE 387	Forficula senegalensis	Mai Mahiu (Kenya)	2007	99.7 ± 0.2a
	ICIPE 655	Soil	Kabuti (Kenya)	2008	98.2 ± 0.8ab
B. b.	ICIPE 10	Soil	Mbita (Kenya)	2002	99.5 ± 0.3a
	ICIPE 273	Soil	Mbita (Kenya)	2006	99.5 ± 0.5a
	ICIPE 279	Coleopteran larvae	Kericho (Kenya)	2005	98.5 ± 0.6ab
	ICIPE 603	Hymenoptera	Taita (Kenya)	2007	90.7 ± 1.5g
	ICIPE 609	Soil	Meru (Kenya)	2008	91.7 ± 2.8fg
	ICIPE 621	Soil	Kericho (Kenya)	2008	95.7 ± 2.8bcd
	ICIPE 644	Soil	Mauritius	2007	93.1 ± 2.7defg
	ICIPE 676	Soil	Muhaka (Kenya)	2008	95.2 ± 2.0cde
Isaria sp.	ICIPE 682	Soil	Mabokoni (Kenya)	2008	99.7 ± 0.2a

* Viability of fungal isolates conidia growing on SDA media determined after 18h at 26±2°C. Means within same column followed by same lower-case letters are not significantly different by Tukey HSD multiple range test at $p < 0.05$. [M.a. = M. anisopliae; B.b. = B. bassiana; I. = Isaria sp.]

Mean mortality in the control ranged between 10-20% seven days after treatment. All the three fungal species were pathogenic to A. craccivora. However, mortality varied significantly among the isolates (F=61.89; df=22; P<0.0001) and varied between 34.5% for the least virulent isolate (M. anisopliae ICIPE 18) and 90.0% for the most virulent isolate

(*M. anisopliae* ICIPE 62) seven days after treatment (Table 3.2). The lethal time mortality (LT_{50}) values were calculated for 21 isolates that caused >50% mortality. They ranged between 3.3 and 6.3 days and were significantly different among isolates (F=8.53; df=20; P<0.0001) (Table 3.2). Based on short LT_{50} values, two isolates of *M. anisopliae* (ICIPE 62 and ICIPE 41) and one isolate of *B. bassiana* (ICIPE 644) were selected for dose-response and conidial production on aphid cadaver bioassays.

Table 3.2: Virulence of *M. anisopliae*, *B. bassiana* and *Isaria* sp isolates against apterous adult *Aphis craccivora*

Fungal species	Fungal isolate	% Mortality ± SE	LT_{50} (days) 95% FL
M. anisopliae	ICIPE 7	44.0 ± 6.0de	-
	ICIPE 18	34.5 ± 3.2e	-
	ICIPE 20	50.5 ± 9.5cde	5.8 (5.7-6.0)
	ICIPE 30	57.7 ± 2.7bcde	5.3 (5.2-5.4)
	ICIPE 40	61.7 ± 6.2bcde	4.6 (4.5-4.7)
	ICIPE 41	79.5 ± 3.1ab	3.6 (3.5-3.7)
	ICIPE 62	90.0 ± 1.2a	3.3 (3.2-3.4)
	ICIPE 63	60.7 ± 4.6bcde	4.8 (4.7-5.0)
	ICIPE 68	72.0 ± 6.0abc	4.5 (4.4-4.6)
	ICIPE 69	71.2 ± 2.6abcd	4.2 (4.1-4.3)
	ICIPE 74	71.2 ± 2.2abcd	4.6 (4.5-4.7)
	ICIPE 78	64.2 ± 2.7abcd	4.8 (4.6-4.9)
	ICIPE 387	64.7 ± 6.2abcd	4.3 (4.2-4.5)
	ICIPE 655	65.7 ± 6.0abcd	4.4 (4.3-4.5)
B. bassiana	ICIPE 10	50.0 ± 2.4cde	6.3 (6.2-6.5)
	ICIPE 273	57.5 ± 3.2bcde	5.5 (5.3-5.7)
	ICIPE 279	63.0 ± 2.6abcd	4.8 (4.7-5.0)
	ICIPE 603	69.0 ± 1.7abcd	4.8 (4.7-4.9)
	ICIPE 609	67.5 ± 4.7abcd	4.3 (4.2-4.4)
	ICIPE 621	70.5 ± 1.5abcd	4.4 (4.3-4.5)
	ICIPE 644	74.7 ± 5.0ab	3.7 (3.6-3.8)
	ICIPE 676	54.2 ± 8.0bcde	5.9 (5.7-6.0)
Isaria sp.	ICIPE 682	64.2 ± 12.2abcd	4.5 (4.2-4.8)

Mean within a column followed by same lower-case letters are not significantly different by Tukey's HSD multiple range test at p<0.05. Lethal time 50 (LT_{50}) (days) ± 95% fiducial limit (FL).

3.4.2 Dose-mortality response

The lethal concentration (LC_{50}) values varied between the fungal isolates with ICIPE 62 having the lowest values (2.3×10^6 conidia ml^{-1}), followed by ICIPE 41 (1.3×10^8 conidia ml^{-1}) and ICIPE 644 (1×10^9 conidia ml^{-1}) seven days post-treatment (Figure 3.1). Computed relative potency ratios ranged from 0.002 to 1.00, and comparison between the three isolates revealed that ICIPE 62 was most potent than the two other isolates (Figure 3.1).

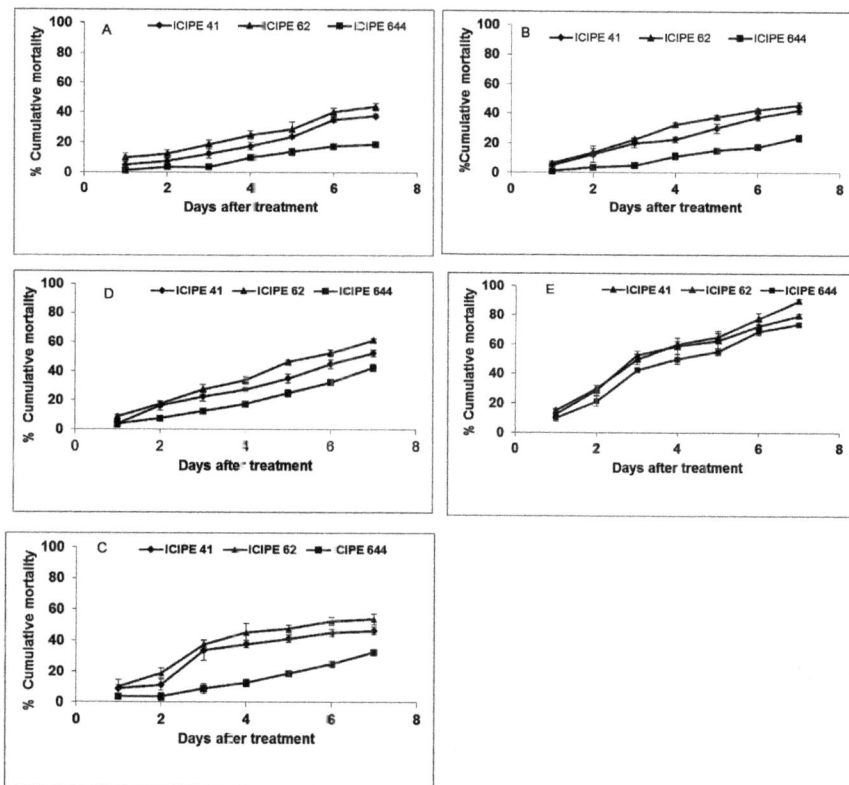

Figure 3.1: Dose dependent mortality response over time on *Aphis craccivora* induced by ICIPE 41, ICIPE 62 and ICIPE 544 from day 1 to 7 days 7 after post-treatment. A-E represents 1×10^4, 1×10^5, 1×10^6, 1×10^7 and 1×10^8 conidia ml^{-1} in that order.

3.4.3 Production of conidia on aphid cadavers

There was significant variation in the number of conidia produced among the three isolates and across days (F=50.59; df=8; P<0.0001) (Figure 3.2). Conidia produced by the isolates were 2.1×10^7, 3.0×10^7 and 1.6×10^7 for ICIPE 41, ICIPE 62 and ICIPE 644, respectively, at three days post-treatment. Conidia production peaked at six days post-treatment with 2.7×10^7, 4.5×10^7 and 2.1×10^7 for ICIPE 41, ICIPE 62 and ICIPE 644 respectively, but dropped at nine days post-treatment with 1.5×10^7, 2.2×10^7 and 1.0×10^7 for ICIPE 41, ICIPE 62 and ICIPE 644 respectively. However, ICIPE 62 produced the highest number of spores across all the days as compared to the two other isolates (Figure 3.2).

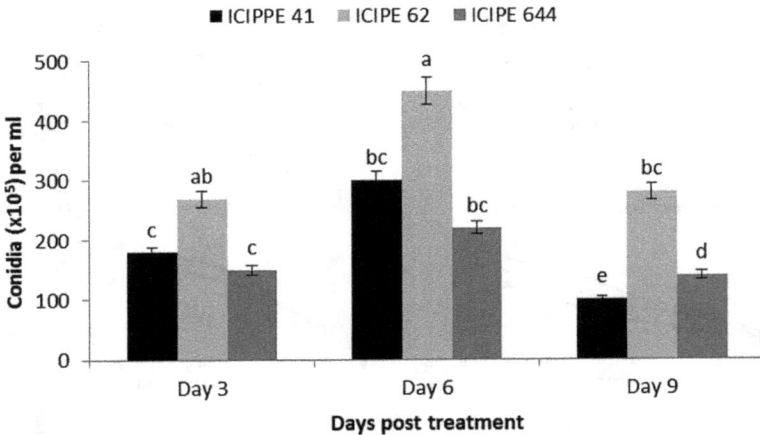

Figure 3.2: Conidial production on individual adult apterous *Aphis craccivora* treated with fungal concentrations of 1×10^8 conidia ml^{-1} at 3, 6, and 9 days post-treatment. Data presented are mean ± SE at p<0.05

3.4.5 Efficacy of aqueous and emulsifiable formulations of *Metarhizium anisopliae* ICIPE 62 against *A. craccivora* in screenhouse

The initial aphid populations before the application of treatments was 333.8 ± 30.6 and 322 ± 31.9, for emulsifiable and aqueous formulations while control population was 350.2 ± 33.2 and 305.1 ± 28.9 for oil and water, respectively. Following application of the treatments, the number of aphids on plants was significantly reduced (F=49.42; df=8; P<0.0001) in both aqueous and emulsifiable formulations (Figure 3.3).

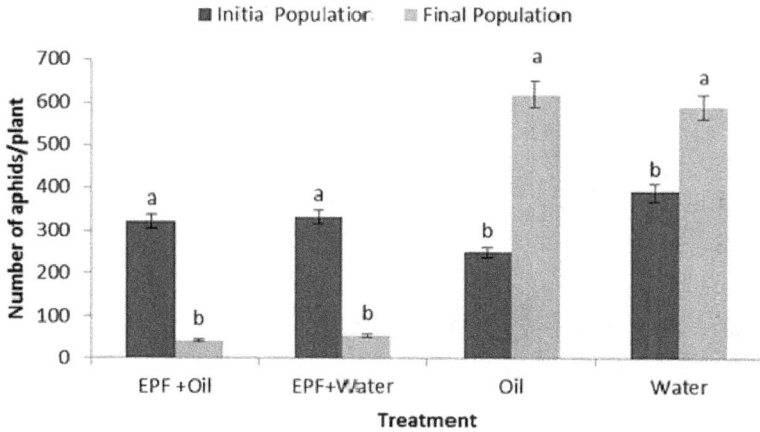

Figure 3.3: Mean number of *Aphis craccivora* per plant following application of aqueous (FA) and emulsifiable (FE) formulations of *Metarhizium anisopliae*. Controls were treated with aqueous (CA) and emulsifiable (CE) formulations without fungal conidia. Bars represent means ± SE, and means followed by same letter are not significantly different by Tukey's HSD multiple range test at p<0.05.

Mortality of *A. craccivora* collected from treated plants in each cage and maintained in the lab was 16 and 18.2% in control (water and oil without conidia in that order) and 80 and 86% in conidia formulated in aqueous and emulsifiable formulation, respectively (Figure 3.4).

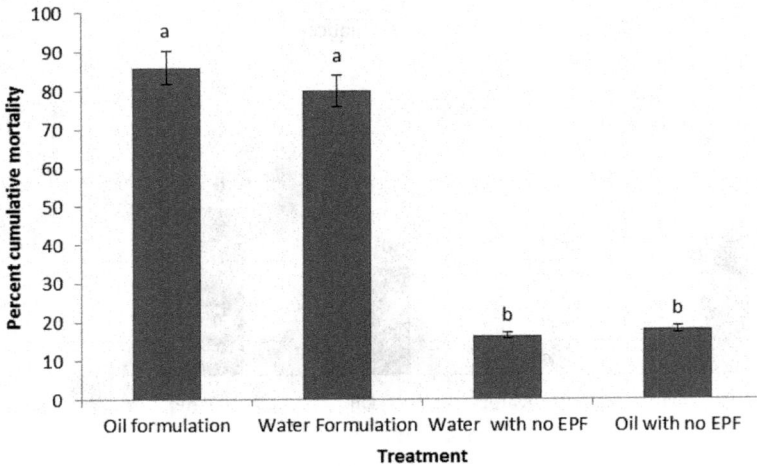

Figure 3.4: Percent mortality of adult apterous *Aphis craccivora* 7 days after treatment with fungal conidia formulated as oil and aqueous, water and corn oil without conidia represent control treatments. Error bars denote means and SE at 95% CI. Means followed by same letter are not significantly different by Tukey's HSD multiple range test at p<0.05

There was no significant difference in aphid population growth rate (r_i) between the aqueous and emulsifiable formulations (t=1.27; df=8; P=0.23) (Figure 3. 5).

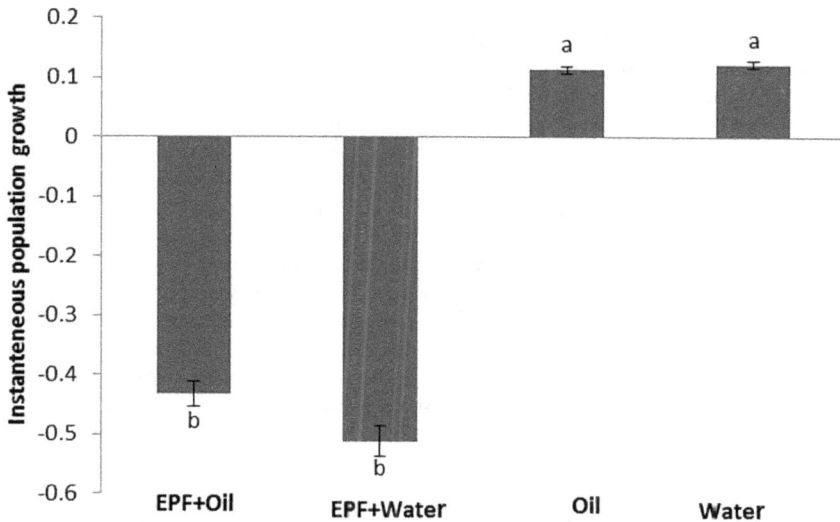

Figure 3.5: Instantaneous growth of *Aphis craccivora* population 7 days post-treatment following application of aqueous (AF) and emulsifiable (EF) formulations of *Metarhizium anisopliae* ICIPE 62. Controls were treated with aqueous (CA) and emulsifiable (CE) formulations without fungal conidia. Bars indicate means and SE at 95% CI. Means followed by the same letter are not significantly different by Tukey's HSD multiple range test at P<0.05

3.5 Discussion

The results of the screening in the present study showed that all the fungal isolates were pathogenic to *A. craccivora*. There was, however, considerable variation in pathogenic activity among the fungal species and fungal isolates as illustrated by the LT_{50} values. Variation in pathogenic activity of *M. anisopliae* and *B. bassiana* isolates against *A. craccivora* has previously been reported. For example, Ekesi et al., (2000) reported mortality of *A. craccivora* ranging between 58-91% and between 66-100% by isolates of *B. bassiana* and *M. anisopliae,* respectively. On the other hand, Saranya et al., (2010) reported 96% and 80% mortality of *A. craccivora* by isolates of *B. bassiana* and *M. anisopliae,* respectively, seven days post-treatment. *Metarhizium anisopliae* isolate ICIPE 62 which outperformed the other isolates in terms of shortest LT_{50}, lowest LC_{50} and highest production of spores on aphid cadavers has previously been reported to be virulent against *A. gossypii, Br. brassicae* and *L. pesudobrassiace* (Bayissa et al., 2016a). Isolate ICIPE 41 (*M. anisopliae*) was the second-best performing isolate in terms of induced lethal time (LT_{50}) and although no previous studies reported its pathogenicity against *A. craccivora*, it has been proved to be pathogenic to other arthropods pests (Dimbi et al., 2003; Bungee et al., 2009). The isolate ICIPE 62 has therefore a broad host range activity which is an advantage for its development as mycoinsecticide. Bayissa et al., (2016b) have demonstrated that, ICIPE 62 is less pathogenic to *Cheilomenes lunata* (Fabricus; Coleoptera: Coccinellidae) a predator of aphids under laboratory conditions. However further studies are warranted to assess the non-target effects of this potent fungal biopesticide on other beneficial arthropods before its application at the field level. A mycoinsecticide based on the isolate ICIPE 62 would be used to control different species of aphids since its efficacy on different aphid species has been reported (Bayissa et al., 2016a). The spore production of ICIPE 62 on aphid cadavers reached its peak at day six but declined on day nine post-treatment. Spore production on insect cadavers is highly dependent on relative humidity but temperature, fungal isolate species, host species and its stage of development as well as incubation period have also been shown to significantly influence spore production (Sosa-Gómez and Alves 2000). Since the cadavers were incubated in the same conditions, we therefore hypothesize that the spore production decline at day nine post-treatment might be due to longer incubation period. High spore production on insect cadavers has been shown to positively influence virulence of fungal pathogens (Mascarin et al., 2013; Niassy et al., 2012). High density of infective spores of a pathogen is required to trigger an epizootic that could contribute to reduction in pest population (Carruthers et al., 1991). The current study evaluated the potential of *M. anisopliae* isolate ICIPE 62 under laboratory conditions and although the isolate is commercialized, further studies are warranted to evaluate its efficacy against *A. craccivora* under different field conditions; since the efficacy of these EPFs is mostly influenced by biotic factors related to the host insect (behavior and genetics), pathogens characteristics (virulence and persistence) after application (Pettersson et al., 1998; Steinkraus 2006; Purandare and Tenhumberg 2012; Lopez-Perez et al., 2015), and abiotic factors including sunlight or UV, rainfall, temperature and humidity (Roy et al., 2006; Eyheraguibel et al., 2010; Jaronski 2010).

Similar results were reported on *M. persicae* with the same isolate by Bayissa et al., 2016a. This represents another advantage of this isolate since sporulated cadavers can act as source of secondary infection, especially for insects with gregarious behavior such as *A. craccivora* (Pettersson et al., 1998; Purandare and Tenhumberg 2012). This is a key consideration in selection of a potential microbial insecticide because the higher the efficacy and the lower the application rate (dose) of a microbial insecticide, the more likely it might be economical and more acceptable to users.

The speed of kill is a key factor when considering selection and commercialization of a microbial insecticide. Although the EPF based mycoinsecticides are characterized by slow killing process, the ability of insect pests to cause damage to crops is reduced towards the terminal days after infection by EPFs (Leng et al., 2011; Mohammadbeigi and Port 2015), which also presents an advantage on the use of microbial insecticides especially when total eradication is not the primary objective (Faria et al., 1999); and where mycoinsecticides are used in an IPM programme.

Commercial use of microbial insecticides depends largely on cost-effective production systems that produce high yields and concentrations of infective fungal propagules or conidia (Jaronski; 2013). Therefore, virulence and relative potency of fungal isolates is an important aspect in selection and commercialization of mycoinsecticides. Evaluation of relative potency in this study revealed that among the best three isolates, *M. anisopliae* isolate ICIPE 62 was 50 times and 500 times more potent than ICIPE 41 (*M. anisopliae*) and ICIPE 644 (*B. bassiana*), respectively. The ability of potential mycoinsecticide to produce spores on dead insects is advantageous since it can act as source of inoculum for secondary infections further reducing insects' population; and it is also cost effective because it increases the probability of natural infection and spread the epizooty among the pest population (Lacey and Siegel 2000; Roy and Pell, 2000). Secondary infection by EPF propagules is aided by host behaviour and *A. craccivora* is gregarious and forms large colonies feeding on young growing tips of plants. This aggregation predisposes them to infection as they are in close contact with each other and their soft body increases susceptibility of infection by EPFs (Pettersson et al., 1998; Purandare and Tenhumberg, 2012).

In the screenhouse experiments, application of aqueous and emulsifiable formulations of *M. anisopliae* isolate ICIPE 62 resulted in significant decline in aphid populations, while increases in the controls were observed. The increase in aphid population in control implies that oil and water without conidia were not toxic to *A. craccivora*. However, there was no significant difference between both formulations. Similar results were reported by Bayissa et al., 2016a, with the same fungal isolate on *Br. brassicae* and *L. pseudobrassicae,* except on *A. gossypii* where emulsifiable formulation performed better. The lack of significant difference between emulsifiable and aqueous formulations could be attributed to the fact that in the screenhouse conidia are not directly exposed to solar radiations which are detrimental to EPF.

3.6 Conclusions and recommendations

In conclusion, this study has identified *M. anisopliae* ICIPE 62 as virulent against *A. craccivora* in terms of LT_{50} and LC_{50} values, and high conidial production on the cadavers, and could therefore be considered for development of mycoinsecticide against this pest on cowpea and other vegetables. The development of this potential EPF based biopesticide could also enhance Aphid-IPM strategy in cowpea production systems. However, considering that different environmental conditions could affect the performance of an entomopathogenic fungus, field evaluation is warranted to confirm its efficacy under different agro-ecologies.

3.7 References

Abate T and Ampofo JKO. 1996. Insect pests of beans in Africa: their ecology and management. *Annual Review of Entomology*, 41 (1):45-73.

Abbott W.A. 1925. Method of computing the effectiveness of an insecticide. *Journal of Economic Entomology*, 18 (2):265–267.

Asante SK, Tamo M and Jackai LEN. 2001. Integrated management of cowpea insect pests using elite cultivars, date of planting and minimum insecticide application. *African Crop Science. Journal*, 9 (4):655–665.

Baidoo PK, Baidoo-Ansah D and Agbonu I. 2012. Effects of Neem (*Azadirachta indica* A. Juss). Products on *Aphis craccivora* and its predator *Harmonia axyridis* on cowpea. *American Journal of Experimental Agriculture*, 2:198-206.

Baidoo PK and Mochiah MB. 2014. Varietal Susceptibility of Improved Cowpea *Vigna unguiculata* (L.) (Walp.) cultivars to field and storage pests. *Sustainable Agriculture Research*, 3 (2):69.

Banken J. 1996. An ecotoxicological assessment of the neem insecticide, Neemix, on the pea aphid (*Acyrthosiphon pisum* Harris) and the seven-spot lady beetle (*Coccinella septempunctata* L.). Doctoral dissertation, Washington State University, USA.

Bayissa W, Ekesi S, Mohamed SA, Kaaya GP, Wagacha JM, Hanna R and Maniania NK. 2016a. Selection of fungal isolates for virulence against three aphid pest species of crucifers and okra. *Journal of. Pest Science*, 90:355-68.

Bayissa W, Ekesi S, Mohamed SA, Kaaya GP, Wagacha JM, Hanna R and Maniania NK. 2016b. Interactions among vegetable-infesting aphids, the fungal pathogen *Metarhizium anisopliae* (Ascomycota: Hypocreales) and the predatory coccinellid *Cheilomenes lunata* (Coleoptera: Coccinellidae). *Biocontrol Science and Technology*, 26 (2):274-290.

Birch LC. 1948. The intrinsic rate of natural increase of an insect population. *The Journal of Animal Ecology*, 17:15–26.

Blackman RL and Eastop VF. 2000. Aphids on the World's Crops. An Identification and Information Guide. 2nd ed. Wiley and Chichester, London, United Kingdom, 466 pp.

Boukar O, Massawe F, Muranaka S, Franco J, Maziya-Dixon B, Singh B and Fatokun C. 2011. Evaluation of cowpea germplasm lines for protein and mineral concentrations in grains. *Plant Genetic Resources*, 9(4):515–522.

Boyetchko S, Pedersen E, Punja Z and Reddy M. 1999. Formulations of Biopesticides. In F.R. Hall, and J.J. Menn (eds). Biopesticides: Use and delivery. Methods in biotechnology. Vol. 5 Humana Press, 487-508 pp.

Bugeme DM, Knapp M, Boga HI, Wanjoya HK and Maniania NK. 2009. Influence of temperature on virulence of fungal isolates of *Metarhizium anisopliae* and *Beauveria bassiana* to the two-spotted spider mite *Tetranychus urticae* Koch. *Mycopathologia*, 167 (4):221-227.

Burgerjon A. 1956. Pulvérisation de poudrage au laboratoire par des préparations pathogè`e`nes insecticides. *Annales des Epiphyties*, 677–688.

Carruthers RI, Sawyer AJ and Hural K. 1991. Use of fungal pathogens for biological control of insect pests. In Sustainable Agriculture Research and Education in the Field. A Proceedings by National Research and Council Board on Agriculture 15 January, 1991. National Academy Press, Washington, USA, 336-372 pp.

Chandler D, Davidson G, Grant WP, Greaves J and Tatchell GM. 2008. Microbial biopesticides for integrated crop management: an assessment of environmental and regulatory sustainability. *Trends in Food Science and Technology*, 19 (5):275-283.

Chandler D, Bailey AS, Tatchell GM, Davidson G, Greaves J and Grant WP. 2011. The development, regulation and use of biopesticides for integrated pest management. Philosophical Transactions of the royal society of London. Series B, *Biological Sciences*, 366 (1573):1987–98.

Cherry AC and Gwynnn RL. 2007. Perspective on the development of biocontrol in Africa. *Biocontrol Science and. Technology*, 17 (7):665-676.

Dimbi S, Maniania NK, Lux SA, Ekesi S and Mueke JK. 2003. Pathogenicity of *Metarhizium anisopliae* (Metsch.) Sorokin and *Beauveria bassiana* (Balsamo) Vuillemin, to three adult fruit fly species: *Ceratitis capitata* (Weidemann), *C. rosa* var. *fasciventris* Karsch and *C. cosyra* (Walker) (Diptera: Tephritidae). *Mycopatholgia*, 156 (4):375–382.

Ehlers JD and Hall AE.1997. Cowpea (*Vigna unguiculata* L. Walp.). *Field Crops Research*, 53 (1-3):187-204.

Ekesi S, Akpa AD, Onul and Ogunlana MO. 2000. Entomopathogenicity of *Beauveria bassiana* and *Metarhizium anisopliae* to the cowpea aphid, *Aphis craccivora* Koch (Homoptera: Aphididae). *Archieves of Phytopathology and Plant Protection*, 33 (2):171–180.

Eyheraguibel B, Richard C, Ledoigt G and Ter Halle A. 2010. Photoprotection by plant extracts: a new ecological means to reduce pesticide photodegradation. *Journal of Agricultural and Food Chemistry*, 58 (17):9692-9696.

Faria MRD, Almeida DDO and Magalhães BP. 1999. Food consumption of *Rhammatocerus schistocercoides* Rehn (Orthoptera: Acrididae) infected by the fungus *Metarhizium flavoviride* Gams & Rozsypal. *Anais da Sociedade Entomológica do Brasil*, 28 (1):91–99.

Faria MRD and Wraight SP. 2007. Mycoinsecticides and mycoacaricides: a comprehensive list with worldwide coverage and international classification of formulation types. *Biological Control*, 43 (3):237-256.

Fatokun CA. 2000. Breeding cowpea for resistance to insect pests: attempted crosses between cowpea and *Vigna vexillata*. Challenges and Opportunities for Enhancing Sustainable Cowpea Production, pp. 52–61. In C.A. Fatokun, S.A. Tarawali, B.B. Singh, P.M. Kormawa, M. Tamo (eds.) Proceedings of the World Cowpea Conference III 4 September 2000, Ibadan, Nigeria. International Institute of Tropical Agriculture (IITA).

Goettel MS, Inglis GD and Wraight SP. 2000. Fungi. In: A. Lace, and H.K. Kaya (eds), Field manual of techniques in invertebrate pathology. Springer, Dordrecht, Netherlands, 255-282 pp.

Gwynn R. and J.K. Maniania. 2010. Africa with special reference to Kenya, 99 pp. In: J.T Kabaluk., A.M. Svircev, M.S. Goettel, S.G. Woo (eds.), The use and regulation of microbial pesticides in representative jurisdictions worldwide. IOBC global. www.IOBC-Global.org.

Hajek AE and. St. Leger RJ. 1994. Interactions between fungal pathogens and insect hosts. *Annual review of entomology*, 39 (1):293-322.

Hall AE. 2004. Breeding for adaptation to drought and heat in cowpea. *European Journal of Agronomy*, 21 (4):447-454.

Hall AE. 2012. Phenotyping cowpeas for adaptation to drought. *Frontiers in Physiology*, 3:1-8 http://realipm.com/index.php/products_cat/bio-pesticides/. Real IPM.

Inglis GD, Enkerli J and Goettel MS. 2012. Laboratory techniques used for entomopathogenic fungi: Hypocreales. In: L.A. Lacey (ed), Manual of techniques in invertebrate pathology 2nd ed. Academic Press, London, UK189-253 pp.

Jackai LEN and Asante SK. 2003. A case for the standardization of protocols used in screening cowpea, *Vigna unguiculata* for resistance to *Callosobruchus maculatus* (Fabricius) (Coleoptera: Bruchidae). *Journal of Stored Products Research*, 39 (3) 251-263.

Jaronski ST. 2010. Ecological factors in the inundative use of fungal entomopathogens. *BioControl*, 55 (1):159-185.

Jaronski ST. 2013. Mass production of entomopathogenic fungi-state of the art. Mass production of beneficial organisms: invertebrates and entomopathogens. Academic, San Diego, 357-413 pp.

Jenkins NE, Heviefo G, Langewald J, Cherry AJ and Lomer CJ. 1998. Development of mass production technology for aerial conidia for use as mycopesticides. *Biocontrol News and Information*, 19:21-32.

Lacey LA and Siegel JP. 2000. Safety and ecotoxicology of entomopathogenic bacteria. In J. F. Charles, A. Delecluse and C. Nielsen-LeRoux (eds.), Entomopathogenic bacteria: from laboratory to field application, Kluwer Academic Press: Dordrecht, The Netherlands, 253-273 pp.

Lacey LA, R. Frutos R, Kaya HK and Vail P. 2001. Insect pathogens as biological control agents: do they have a future? *Biological Control*, 21 (3):230-248.

Leng P, Zhang Z, Pan G and. Zhao M. 2011. Applications and development trends in biopesticides. *African Journal of Biotechnology*, 10 (86):19864-19873.

Lopez-Perez M, Rodriguez-Gomez D and Loera O. 2015. Production of conidia of *Beauveria bassiana* in solid-state culture: current status and future perspectives *Critical Reviews in Biotechnology*, 35 (3):334-341.

Maniania NK. 1993. Evaluation of three formulations of *Beauveria bassiana* (Bals.) Vuill. For control of the stem borer *Chilo partellus* (Swinhoe) (Lep. Pyralidae). *Journal of Applied Entomology*, 115 (1-5):266-272.

Mascarin GM, Kobori NN, Quintela ED and Delalibera I. 2013. The virulence of entomopathogenic fungi against *Bemisia tabaci* biotype B (Hemiptera: Aleyrodidae) and their conidial production using solid substrate fermentation. *Biological control*, 66(3):209-218.

Mohammadbeigi A and Port G. 2015. Effect of infection by *Beauveria bassiana* and *Metarhizium anisopliae* on the feeding of *Uvaroristia zebra*. *Journal of Insect Science*, 15(1):88.

Montesinos E. 2003. Development, registration and commercialization of microbial pesticides for plant protection. *International Microbiology*, 6(4):245-252.

Muchero W, Diop NN and Bhat PRA 2009. Consensus genetic map of cowpea [*Vigna unguiculata* (L) Walp.] and synteny based on EST-derived SNPs, pp 18159–18164. In proceedings of the National Academy of Sciences 27 October, 2009, publication 43. National. Academy of Sciences, U S A.

Nampala P, Kyamanywa S, Ogenga-Latigo MW, Adipala E, Karungi J, Oyobo N, Obuo JE and Jackai LEN. 1999. Integrated management of major field pests of cowpea in Eastern Uganda. *African Crop Science* Journal, 7 (74):479-486.

Niassy S, Maniania N, Subramanian S, Gitonga L, Mburu D, Masiga D and Ekesi S. 2012. Selection of promising fungal biological control agent of the western flower thrips *Frankliniella occidentalis* (Pergande). *Letters in Applied Microbiology*, 54 (6):487-493.

Obopile M. 2006. Economic threshold and injury levels for control of cowpea aphid, *Aphis craccivora* Linnaeus (Homoptera: Aphididae) on cowpea. *African Plant Protection*, 12(1): 111-115

Ofuya TI. 1997. Control of the bean aphid *Aphis craccivora* Koch (Homoptera: Aphididae) in cowpea, *Vigna unguiculata* (L.) Walp. *Integrated pest management reviews*, 2 (4):199-207.

Omongo CA, Ogenga-Latigo MW, Kyamanywa S and Adipala E. 1997. The effect of seasons and cropping systems on the occurrence of cowpea pests in Uganda. In: African Crop Science Conference Proceedings, vol. 3.

Pettersson J, Karunaratne S, Ahmed E and Kumar V. 1998. The cowpea aphid, *Aphis craccivora*, host plant odours and pheromones. *Entomologia Experimentalis et Applicata*, 88 (2):177-184.

Purandare SR and Tenhumberg B. 2012. Influence of aphid honeydew on the foraging behaviour of *Hippodamia convergens* larvae. *Ecological Entomology*, 37 (3):184-192.

R Core 2013. Team: A language and environment for statistical computing. R Foundation for Statistical Computing, Vienna, Austria. URL http://www.R-project.org/.

Roy HE and Pell JK. 2000. Interactions between entomopathogenic fungi and other natural enemies: implications for biological control. *Biocontrol Science and Technology*, 10 (6):737-752.

Roy HE, Steinkraus DC, Eilenberg J, Hajek AE and Pell JK. 2006. Bizarre interactions and endgames: entomopathogenic fungi and their arthropod hosts. *Annual Review of Entomology*, 51:331-357.

Rusoke DG and Rubaihayo PR. 1994. The influence of some crop protection management practices on yield stability of cowpeas. *African Crop Science Journal*, 2 (1):43-48.

Sahayaraj K and Borgio JF. 2010. Virulence of entomopathogenic fungus *Metarhizium anisopliae* (Metsch.) Sorokin on seven insect pests. *Indian Journal of Agricultural Research*, 44 (3):195-200.

Saidi M, Itulya FM, Aguyoh JN, Mshenga PM and Owour G. 2010. Effects of cowpea leaf harvesting initiation time on yields and profitability of a dual-purpose sole cowpea and cowpea-maize intercrop. *Electronic Journal of Environmental, Agricultural and Food Chemistry*, 9 (6):1134-1144.

Saranya S, Ushakumari R, Jacob S and Philip BM. 2010. Efficacy of different entomopathogenic fungi against cowpea aphid, *Aphis craccivora* (Koch). *Journal of Biopesticides*, 3 (Special Issue) 138-142.

Shapiro SS and Wilk MB.1965. An analysis of variance test for normality. *Biometrika*, 53 (2):591–611.

Singh BB and Sharma B. 1996. Restructuring cowpea for higher yield. *The Indian Journal of Genetics and Plant Breeding,* 56 (4):389-405.

Singh BB, Ehlers JD, Sharma B and Freire Filho FR. 2000. Recent progress in cowpea breeding. Challenges and Opportunities for Sustainable Cowpea Production pp. 22-40. In C.A. Fatokun, S.A. Tarawali, B.B. Singh, P.M. Kormawa and M. Tamo (eds.), proceedings of the World Conference III. 4 September 2000, Ibadan, Nigeria. International Institute of Tropical Agriculture (IITA).

Sosa-Gómez DR and Alves SB. 2000. Temperature and relative humidity requirements for conidiogenesis of *Beauveria bassiana* (Deuteromycetes: Moniliaceae). *Anais da Sociedade Entomologica do Brasil*, 29 (3):515–521.

Steinkraus DC. 2006. Factors affecting transmission of fungal pathogens of aphids. *Journal of invertebrate pathology*, 92 (3):125-131.

Vu VH, Hong S and Kim K. 2007. Selection of pathogenic fungi for aphid control. *Journal of Bioscience and Bioengineering.* 104 (6):498-505.

4.0 EFFICACY OF AQUEOUS AND OIL FORMULATIONS OF *METARHIZIUM ANISOPLIAE* ISOLATE AGAINST *APHIS CRACCIVORA* UNDER FIELD CONDITIONS

Submitted to Journal of Applied Entomology as: Efficacy of aqueous and oil formulations of *Metarhizium anisopliae* isolate against *Aphis craccivora* under field conditions.

Mweke A, Ulrichs C, Akutse KS, Fiaboe KKM, Nana P, Maniania NK and Ekesi S

4.1 Abstract

Cowpea (*Vigna unguiculata* L. Walp) production is constrained by biotic and abiotic factors, among which Cowpea aphid (*Aphis craccivora* Koch) is ranked a key insect pest that severely limits its potential for provision of food and nutritional security to millions of people in Sub-Saharan Africa (SSA). The use of entomopathogenic fungi (EPF) for the management of *A. craccivora* has been recently demonstrated at laboratory and field levels as alternative to synthetic insecticides, however their use in Africa is still low. This study was designed to assess the efficacy of aqueous and oil formulations of *Metarhizium anisopliae* isolate ICIPE 62 against *A. craccivora* under field conditions. EPF formulations (1×10^{12} conidia ml^{-1}) and a commonly used insecticide-Lambda-cyhalothrin (Duduthrin (1.75 g AI) were applied using knapsack sprayers with target output of 350 L ha^{-1}. Data on aphid infestation levels were collected weekly. Efficacy of EPF in inducing mortality was also assessed 24 hours post-treatment by collecting aphids from treated plots and assessing mycosis in dead aphids. After treatment application for 8 weeks in the first season characterized by heavy and frequent rains, there was no significant reduction in aphid population density in fungus-treated plots compared to the control and Duduthrin treated plots. However, in the second season which was dry and hot, 6 weeks after applying the treatments, oil formulation application resulted in low aphid density compared to control and Duduthrin treated plots. EPF formulations did not negatively affect the natural enemies' population. Application of treatments in season 1 did not confer Leaf yield benefits but the two fungal formulations recorded higher leaf yields in season two compared to other treatments. Grain yield in season one was lower in control and Duduthrin treated plots compared to the two EPF formulations, while in season two oil formulation produced more grain yield compared to control. This study showed that both aqueous and oil formulations of *M. anisopliae* isolate ICIPE 62 are effective in suppressing *A. craccivora* population under field conditions without adverse effects on its beneficial insects.

Keywords: Cowpea, Entomopathogenic fungi, ICIPE 62, suppression, management

4.2 Introduction

African leafy vegetables including cowpea form an important component of diet and provide income to smallholder farmers in rural areas as well and urban and peri-urban dwellers (Abukutsa- Onyango, 2010; Rusike et al., 2013). Cowpea is well adapted in the tropics because it tolerates dry weather conditions and does well even in poor soils (Mucheru-Muna, 2010; Bisikwa et al., 2014; Ddamulira et al., 2015). Cowpea leaves are rich in proteins, carbohydrates, minerals and vitamins and have medicinal attributes (Hall, 2012; Trehan et al., 2015).

Despite the nutritional and income generation potential of the crops to millions of people worldwide, its production has mainly been hampered by arthropod pests and diseases (Jackai and Daoust, 1986; Dugje et al., 2009). Cowpea aphid *(Aphis craccivora* Koch) is a sap sacking soft insect that attacks cowpea from seedling to podding stages, and also a vector of viral diseases to the host plant. It voraciously feeds on soft and actively growing plant tissues and directly impacts the crop by removing vital plant sap and interfering with photosynthetic functions that negatively affect the productivity or yield of the crop (Blackman and Eastop 2006; Souleymane et al., 2013).

Chemical pesticides are extensively used to manage *A. craccivora* partly because of its susceptibility to different groups of pesticides, and also because it is not feasible to grow cowpea profitably without use of pesticides (Waddington et al., 2010; Egho and Enujeke, 2012). However, the use of theses synthetic chemical pesticides raises concerns about environmental pollution such as contamination of soil and ground water, adverse effect on non-target species and accumulation of residues along the food chain and in the final products. Furthermore, the use of synthetic pesticides in cowpea as a leaf vegetable leads to lose of yield and income due to observation of long post-harvest intervals since cowpea leaves are regularly harvested in a very short period of time (Mweke et al., 2016). These detrimental effects have therefore led to search for alternative strategies for the management of the pest (Ofuya, 1997; Asante et al., 2001; Nampala et al., 2002).

Lambda-cyhalothrin is a contact synthetic pyrethroid insecticide that induces mortality on insects by disrupting the normal functions of central nervous systems and is used in the management of sucking pests including aphids. In Kenya it is commonly used in management of aphids and is sold under different trade names and formulations e.g. Duduthrin ®1.75EC (Twiga Chemical Industries Ltd), KARATE® 2.5WG, KARATE® 5SC (Syngenta East Africa Ltd), LAMBDEX 5 EC (Amiran Kenya Ltd) among others. This insecticide has a known persistence in soil and has been detected in produce even when post-harvest intervals have been observed (Hornsby et al., 1995; Kithure et al., 2017).

Aphids are susceptible to infection by micro-organisms including entomopathogenic fungi (EPF) that are responsible for regulation of their populations under natural environment (Feng et al., 1990; Hajek and St. Leger, 1994; Shah and Pell, 2003; St. Leger and Wang, 2009; Selvaraj et al., 2010). Commercial biopesticides derived from entomopathogenic fungi in Zygomycetes and Hyphomycetes groups, mainly *Lecanicillium lecanii* (Zimm.) Viegas, *Metarhizium anisopliae*

(Metsch.) Sorokin, *Beauveria bassiana* (Bals.) Vuill. and *Isaria* spp., have been successfully used in the management of several insect pests including different aphid species. These EPF are cheap and easy to mass-produce on organic substrates like rice, maize and sorghum and produce stable infective spores (Milner, 1997 Jaronski, 2013). These biopesticides have advantages of wide host range infections and are compatible with integrated management interventions (Milner, 1997; Lacey et al., 2001; Lacey, 2015). However, adoption and application of EPF based biopesticides is hampered by their short shelf life, inconsistent performance and low persistence under field conditions.

Different fungal based biopesticides formulations have been used to improve these undesirable characteristics and improve market share and performance of these products under different environmental conditions (Inglis et al., 2002; Chandler et al., 2011; Guinossi et al., 2012; Gašić and Tanović, 2013). The EPF *M. anisopliae* isolate ICIPE 62 was previously demonstrated to have significantly high mortality effect on *A. craccivora* under laboratory conditions (Mweke et al., 2018). Formulation of EPF and application method significantly influences their efficacy in pest management under field conditions (Inglis et al., 2002). Additionally, distribution of infective spores has been shown to be influenced by formulation for example conidia formulated in oil are usually evenly distributed on the insect cuticle and leaf surface where else aqueous spore formulations remain as drops on the surface after application and thereby enhancing the efficacy of EPF based biopesticides (Inglis et al., 2000; Inyang et al., 2000; Wraight et al., 2016).

This study was therefore undertaken to evaluate field efficacy of *M. anisopliae* isolate ICIPE 62 propagules formulated as aqueous and oil in the management of *A. craccivora* and compare its performance against commonly used synthetic pyrethroid-Duduthrin under field conditions. The effect of ICIPE 62 on non-target beneficial organisms was also evaluated.

4.3 Materials and methods

4.3.1 Fungal culture and inoculum preparation
Fungal isolate ICIPE 62 (*Metarhizium anisopliae*) used in this study was obtained from *icipe*'s Arthropod Germplasm Centre, and was recently demonstrated to be pathogenic against *A. craccivora* under laboratory conditions (Mweke et al., 2018).

ICIPE 62 conidia were mass produced on sterilized whole long grain rice substrate in Milner bags (60 x 35 cm). The substrate was first autoclaved for 1 hour at 120°C then transferred to plastic buckets 35 (Ø) × 25 (width) ×15 cm (depth) and allowed to cool at room temperature, after which it was inoculated with a 3-day-old culture of blastopores (50 ml) and covered with sterile polyethylene bag. The culture was then incubated for 21 days at ambient conditions (20-26°C, 40-70% RH) (Maniania, 1993). After 21 days, the bag was removed and dried for 5 days at room temperature. Conidia were harvested by sifting the substrate through a sieve (295-μm mesh size) and stored in a refrigerator (4-6°C, 40-50% RH) for less than two weeks before being used in field experiments. Viability of the fungus was determined prior to treatment application by spread plating 100 μL of conidial at a concentration of 3×10^6 conidia ml^{-1} in Sabouraud dextrose agar (SDA) plates. Afterwards the plates were incubated at $26 \pm 2 \circ C$ in darkness and examined after 18 h. The percentage of germination was determined by counting randomly 100 selected conidia a surface area $spot^{-1}$ covered by cover slip under a light microscope (400x) (Goettel and Inglis 1997). The conidia were scored as viable or germinated when the germ tubes were at least as long as the twice the diameter of the conidium (Schumacher and Poehling, 2012). Conidial germination >90% after 18 h on SDA and was considered adequate for use in the field trials. Conidia concentration per gram was calculated by dissolving 0.1 gram of conidia in 10ml of Triton water that had been autoclaved and then serial dilution done to × 100, vortexed and 1 ml pipetted into a hemocytometer and spores counted under the microscope. The number of spores in 0.1ml was used to calculate the amount of conidia in grams required to produce a concentration of 1×10^{12}.

4.3.2 Lambda-cyhalothrin (Duduthrin® l.75EC)
This pesticide was periodically bought from local agro-input suppliers after confirmation of its authenticity and validity by expiry date. The pesticide was prepared by mixing 65ml of the pesticide with 20 litres of clean water in a knapsack sprayer and adding 0.05% (1ml) Integra (sticker, Greenlife Crop Protection Africa Ltd) and shaking thoroughly to produce a homogenous mixture before application.

4.3.3 Experimental sites
The experiment was carried out at *icipe*'s Mbita point campus, Homabay County, Western Kenya for two seasons. In the first season the experiment was carried out between January and April 2016 in a field located at 00.4305S, 034.2057 E, 1150 meters above sea level (m.s.l). The mean annual rainfall in season one between January and April was 120.4mm while the minimum and maximum temperatures were 23°C and 29.2 °C respectively with relative humidity ranging between 60 and 70%. In the second season the experiments were done between May and August 2016 in field and the mean rainfall was 53.67mm. Minimum and maximum temperatures were 20.7 °C and 28.5 °C and relative humidity ranged between 60 and 65% (Kenya Meteorological department, 2017; ICIPE, 2016).

4.3.4 Crop
The land was first ploughed and harrowed in preparation for planting. Ex-Luanda with a known susceptibility to *A. craccivora* obtained from *icipe*'s germplasm collection was used in the experiment. Four cowpea seeds per hill were sown in 10 ×10 m plots at a spacing of 20 cm intra-row by 75 cm inter-row and later thinned to 2 plants per hill 14 days after emergence giving a plant population of 570 plants per plot and translating to 66,666 plants per hectare. Overhead irrigation (sprinkler) was used for the first 3 weeks to support the crop as it was relatively dry in January during planting and irrigation discontinued after the onset of the rains. Weeding was done

twice a month before the crop established and smothered the weeds and the frequency of weeding reduced to once a month. The crop was then left for natural aphid infestation. Cowpea land race Ex-Luanda with a known susceptibility to *A. craccivora* was used in this study.

4.3.5 Treatments, layout and design

The experiment was done for 2 seasons and each season, one experimental field was planted. The experiment had four treatments as follows: (i) an aqueous and (ii) oil formulations of *M. anisopliae* ICIPE 62, (iii) Lambda-cyhalothrin (Duduthrin® 1.75EC) and (iv) the control. In aqueous formulation, spores were suspended in water containing 0.05% Integra (sticker, Greenlife Crop Protection Africa Ltd) with 0.1% nutrient agar, 0.1% glycerin and 0.5% molasses added as protectants and attractant respectively (Maniania, 1993) whereas in oil formulation, spores were suspended in vegetable oil-e ianto (Elianto, Bidco Africa Ltd) with similar proportions of the sticker, nutrient agar, glycerin and molasses as described above in aqueous formulation. The insecticide Duduthrin® 1.75EC (Twiga Chemical Industries Ltd) was applied at the rate of 1.75g (Active Ingredient (AI)) ha^{-1} with 0.1% Integra. Control treatment were sprayed with water containing 0.05% Integra, 0 1% nutrient agar, 0.05% molasses and 0.1% glycerin without any fungal conidia and any insecticide solution. Conidia were applied at the rate of 1 x 10^{12} conidia ml^{-1}. The treatment applications started on day 56 after planting due to late aphids' infestation and thereafter done on a weekly basis for a period of 6 weeks. The fungus formulations and the insecticide were applied with different knapsack sprayers with target output of 350 L ha^{-1} and spraying was done late in the evening between 17:00 and 18:00 h. The experimental design was randomized block design with four replications.

4.3.6 Evaluation of treatments

Two leaflets, each from the base and the top from 20 randomly selected cowpea plants in the middle rows were sampled from each plot for aphid infestation assessment. The aphids were dislodged from the host plant with a fine hair brush into a vial containing 70% alcohol, labelled and thereafter counted in the laboratory. Sampling was done on weekly basis from day 7 after planting until the cowpea leaves begun to dry up.

4.3.7 Asessment of leaf damage

Leaf damage (leaf quality and fitness for human consumption) were visually assessed using the following scale (Benchasri, 2009) 0= visual damage on leaves and flower buds < 10%, 1 = visual damage on leaves and flower buds 10-25%; 2 = visual damage on leaves and flower buds 26-50%; 3 = visual damage on leaves and flower buds 51-75%; 4 = visual damage on leaves and flower buds 76-100%.

4.3.8 Aphid mortality Assessment

Mortality due to the fungus was assessed by actively picking 30 aphids from each fungus treated plot and placing them in plastic dishes (11.3 cm (Ø) × 4 cm (depth) lined with moist filter paper containing sterilized cowpea leaves serving as food source. The lids of the plastic dishes were covered by muslin cloth with apertures (300 x 300 μ) to allow for free circulation of air. The dishes were kept at room temperature and mortality recorded daily for 7 days. The leaves in the dishes were changed daily with fresh ones. Dead aphids were collected and placed in petri-dishes with sterilized moistened filter papers and kept at room temperature for assessing mycosis. Mortality due to the fungus was confirmed by observing mycosis on dead aphid under a dissecting microscope.

4.3.9 Leaf vegetable and grain yield assessment

Leafy vegetable yield data was collected at 7 days interval starting from day 21 after planting (Saidi et al., 2010). The total leaf vegetable weight for each treatment was calculated by cumulatively adding the fresh leaf weights obtained per treatment at the different leaf harvesting dates, and expressed in kilograms per hectare (kg ha^{-1}). At plant maturity the pods were picked, sun dried, threshed and the grain weight recorded using electronic balance and yield computed.

4.3.10 Assessment of natural enemies of *A. craccivora*

Ladybird beetles, spiders, lacewing and parasitoids were the beneficial arthropods assessed in the study. Apart from parasitoid the other arthropods were assessed by counting their numbers on randomly selected plants in each plot. Parasitoids were assessed by collecting 20 mummies per plot and transferring them to perforated petri-dishes and the number of parasitoids emerging were recorded.

4. 3.11 Statistical analysis

The aphid density per plant, natural enemies count data, percent aphid damage score and Leaf and grain yield data were first log transformed before analysis using linear mixed model and means separated using Tukey HSD and data presented in chi-square values. To evaluate aphid mortality induced by entomopathogenic fungi ICIPE 62 formulations; data on mortality were corrected for natural mortality (Abbott, 1925) then, tested for normality test (Shapiro and Wilk, 1965), arcsine transformed then subjected to ANOVA and means separated using Tukey HSD. Data on mycosis were tested for normality test (Shapiro and Wilk, 1965), arcsine transformed and analyzed using ANOVA and means separated using Tukey HSD. Two sample t-test was used to compare performance of the aqueous and oil formulations. Data were analyzed using R software (R Core, 2013).

4.4 Results

4.4.1 Aphid infestation

In season one (with heavy rainfall and cooler temperatures), there was no significant difference between the treatments ($\chi^2_{3, 48}$=4.2; P=0.36) with regard to aphid population density per plant after 8 weeks of treatment evaluation (Table 4.1). (Table 4.1).–In the second season (dry and warm) the mean aphid infestation per plant after 6 weeks of treatment was significantly different between the treatments ($\chi^2_{3,48}$ = 8.2, P= 0.04).

Table 4.1: Mean aphid population density per plant eight after treatment in season 1 and 2

Treatments	Aphid density per plant
Season 1 (wet)	
Oil formulation	2.1 ± 1.4a
Water formulation	3.9 ± 2.1 a
Duduthrin	4.6 ± 2.4a
Control	2.2 ± 1.2 a
Season 2 (dry)	
Oil formulation	1.0 ± 0.8b
Water formulation	1.4 ± 1.4b
Duduthrin	6.9 ± 5.7a
Control	4.0 ± 1.4a

Means within same column followed by same letter are not significantly different by Tukey HSD for each season.

4.4.2 Aphid mortality and infection by EPF

In season 1 mean mortality of aphids induced by EPF was higher in oil formulation (74.3% ± 4.1) compared to aqueous formulation (66.58% ± 3.5); however, there was no significant difference between the two formulations in season 1 (t = 1.43, df = 34, P= 0.16) (Table 4.2). Oil and aqueous formulations of *M. anisopliae* isolate ICIPE 62 caused 71.9% ± 3.2 and 64.8% ± 3.5 infection respectively but did not differ significantly (t=1.85, df=34, P=0.07).–In season 2, the mean mortality induced by oil formulation and aqueous formulations did not differ significantly (t = 1.66, df = 34, P= 0.10). In the same season the mycosis due to the 2 formulations did not differ significantly (t=0.55, df=34, P=0.58) (Table 4.2).

Table 4.2: Mean Mortality and mycosis for two formulations

Season 1 (wet)	% Mortality	% Mycosis
Oil formulation	74.33 ± 4.1a	71.9 ±.3.2a
Water formulation	66.58 ± 3.5a	64.8 ± 3.5a
Control	0.00 ± 0.00	0.00 ± 0.00b
Season 2 (dry)		
Oil formulation	80.0 ± 3.7a	78.7 ± 3.1a
Water formulation	77.9 ± 3.8a	75.5 ± 3.4a
Control	0.00 ± 0.00b	0.00 ± 0.00b

Means within same column followed by same letter are not significantly different by Tukey HSD for each season.

4.4.3 Aphid damage on cowpea

In season 1, Aphid damage on plants was higher in untreated control and Duduthrin treated plots compared to the 2 EPF formulations and was significantly different between the treatments (F=3.43, df=3, 0.02), (Figure 4.1). In season 2 there was significant difference in damage the treatments (F=5.31, df=3, P= 0.001), (Figure 4.1). Damage trend was similar to season 1 where untreated control and Duduthrin recorded the highest damage compared to oil and aqueous formulations of the EPF.

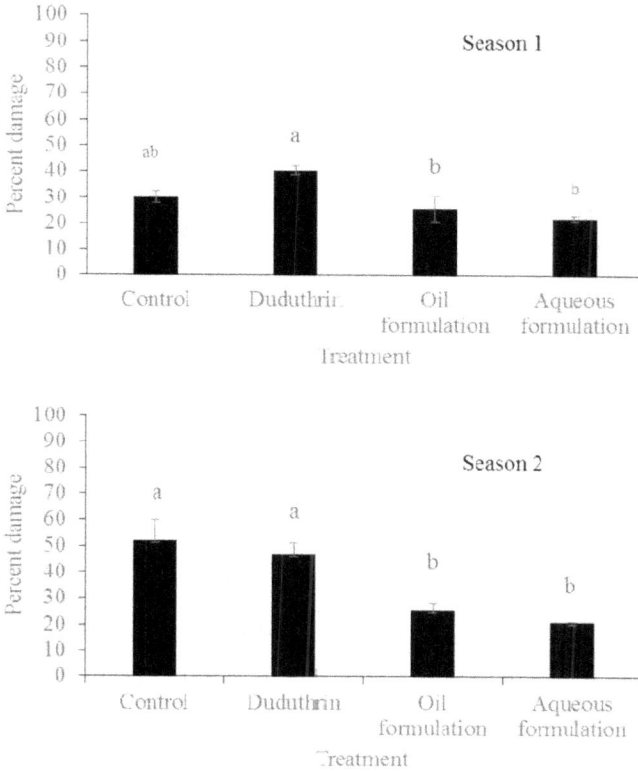

Figure 4.1: Season 1 and 2 aphid damage score. Bars represent means ± SE at p<0.05

4.4.4 Natural enemies of *Aphis cracci vora*

Ladybird beetles, Spiders (*Leucocage decorata*), lacewing and Parasitoid (*Aphidius colemani*) were the natural enemies of *A. cracchora* encountered and identified in the experimental sites. In season 1, among the natural enemies, the spiders (*L. decorata*) were more abundant across the treatments and were significantly lower in plots treated with Duduthrin compared to control and EPF treated plots ($\chi^2_{3,48}$=11.8, P=0.008). The number of ladybird beetles differed significantly among the treatments ($\chi^2_{3, 48}$=28.4, P=0.001) and were higher in EPF treated plots compared to plots treated with Duduthrin and untreated control. The number of lacewings did not differ significantly across the treatments ($\chi^2_{3, 48}$ = 1.8, P=0.61) while the number of parasitoids were the lowest compared to other natural enemies of *A. craccivora* and were also significantly different between the 4 treatments (χ^2 3, 48=12.7, P=0.005), (Table 4. 3).

In the second season the mean number of ladybird beetles varied between the treatments (χ^2 3, 48 =28.2, P=0.001) and were lowest in Duduthrin treated plots. This trend was similar to that in season 1. The mean number of spiders (*L. decorata)* were significantly different between the treatments ($\chi^2$3, 48=24.4 P=0.001) and were lower in Duduthrin treated plots compared to EPF treated plots and untreated control plots while the number of lacewing was significantly different between the treatments (χ^2 3, 48=34.4, P<0.001) and were high in control and EPF-treated plots compared to Duduthrin-treated plots. There was significant difference in parasitoid numbers (*A. colemani*) among the treatments (χ^2 3, 48, =13.9, P=0.001) and were lowest in plots treated with Duduthrin compared to other treatments and their numbers were generally low compared to other natural enemies (Table 4.3).

Table 4.3: Beneficial arthropods density recorded in the various treatments in season 1 and 2

Treatments	Lady bird beetles	Spiders (*Leucocage decorata)*	Green Lacewing	Parasitoid (*Aphidius colemani*)
Season 1 (wet)				
Oil formulation	2.5 ± 1.0a	8.3 ± 2.1a	1.9 ± 0.8a	0.66 ± 0.5a
Water formulation	2.5 ± 1.4a	7.75 ± 1.9a	1.9 ± 0.7a	0.75 ± 0.4a
Duduthrin	1.0 ± 0.5b	6.33 ± 2.2b	1.7 ± 0.8a	0.33 ± 0.2b
Control	1.9 ± 1.1b	9.03 ± 1.9a	1.7 ± 0.6a	0.54 ± 0.4a
Season 2 (dry)				
Oil formulation	7.6 ± 3.4a	8.6 ± 2.8a	4.6 ± 1.9a	1.1 ± 0.7a
Water formulation	6.9 ± 3.4a	4.0 ± 2.6b	4.7 ± 1.9a	1.2 ± 0.5a
Duduthrin	2.3 ± 2.0b	1.3 ± 0.4c	0.8 ± 0.5b	0.37 ± 0.1b
Control	7.2 ± 3.6a	10.6 ± 2.2a	4.7 ± 1.9a	1.0 ± 0.5a

Means within same column followed by same letter are not significantly different by Tukey HSD for each season

4.4.5 Cowpea leaf and grain yield

Cumulative cowpea leaf yield in season 1 did not differ significantly ($\chi^2$3, 48 =0.22, P=0.97) between the treatments (Figure 4.2). However, in season 2 cumulative leaf yield was significantly different ($\chi2$ 3, 48 = 8.0, P = 0.04) between treatments and was higher in the 2 EPF treatments when compared to control and Duduthrin treatments (Figure 4.2).

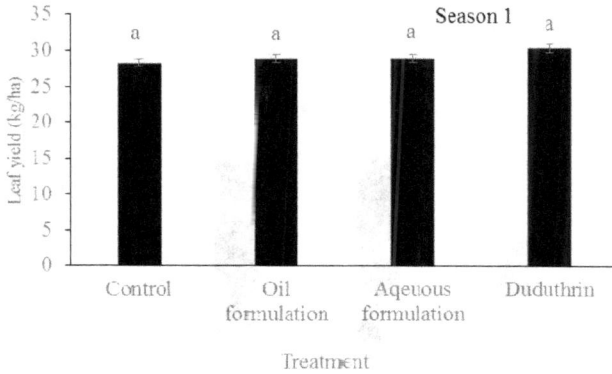

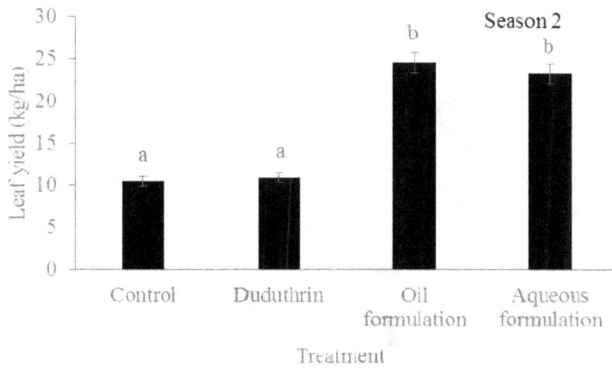

Figure 4. 2: Leaf yield season 1 and 2 in kg ha⁻¹. Bars represent means ± SE at p<0.05

The gain yield was significantly different between treatments in the first season (F=6.65, df=3, P=0.006) where the 2 EFF formulations treatments produced more yield when compared to Duduthrin treatments (Figure 4. 3). In the second season grain yield was higher in oil formulation of the EPF and was significantly different between the 4 treatments (F=3.8, df=3, P=0.03) though was lower when compared to the first season (Figure 4. 3).

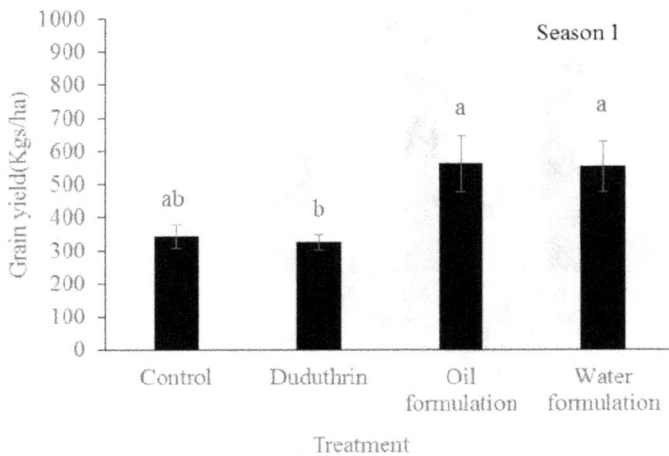

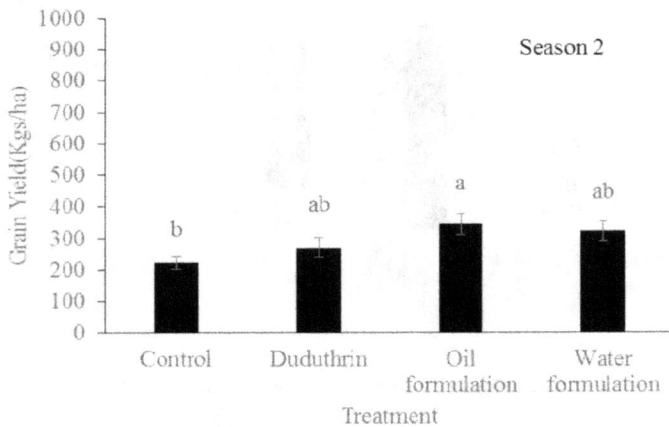

Figure 4.3: Cowpea grain yield season 1 and 2 in kg ha^{-1}. Bars represent means ± SE at p<0.05

4.5 Discussion

Results of this study have demonstrated the potential of *M. anisopliae* isolate ICIPE 62 in management of *A. craccivora* under field conditions. The potential of this isolate to induce high mortality on *A. craccivora* and other aphid species has been demonstrated under laboratory conditions in previous studies (Bayissa et al., 2016; Mweke et al., 2018). Application of EPF formulations in the first season in this study did not result in significant decline in aphid densities. This can be attributed to the fact that the season was characterized by heavy and frequent rains that could have reduced infective amount of the conidia by washing them off the cowpea leave surfaces and aphid colonies thus reducing the levels of infection and subsequent mortality as well as persistence after application in the field. Mode of application (foliar spray) also means that most conidia were dispersed on the leave surface and hence vulnerable to washing off by the rain and irrigation water. Uniform distribution of infective spores on leave surface is an important factor in infection process and the droplet distribution of the spores can be affected by the treatment application method used as well as by the type of formulation (Bateman et al., 1993; Bateman and Alves, 2000; Wraight et al., 2002, Santi et al., 2011; Wraight et al., 2016). Conidia formulated in oil are usually evenly distributed on the insect cuticle and leaf surface whereas aqueous spore formulations remain as drops on the surface after application and thereby enhancing the efficacy of EPF oil based biopesticides formulations (Inglis et al., 2000; Inyang et al., 2000; Wraight et al., 2016). However, this study did report better performance of oil-based formulation compared to aqueous. This could be explained by the fact that efficacy of EPF based biopesticides are not solely influenced by formulation but also by prevailing environmental conditions including rainfall, temperature and relative humidity (Hajek and St. Leger, 1994; Inglis et al., 2002; Nussenbaum et al., 2013).

Rainfall has been shown to negatively influence efficacy of EPF in pest management under field conditions (Fitt et al., 1989; Inglis et al., 2000; Wraight and Ramos, 2002). Infective conidia, deposited on the upper plant leaf surface during treatment application may be washed off by rain or irrigation water where overhead irrigation system is used. Previous studies have shown that rainfall reduces susceptibility of pests to EPF but this is also influenced by the formulation type (Inyang et al., 2000).

Leaf growth leading to expansion of the leaf surface has also been shown to dilute infective spores (Inyang et al., 1998). Though oil formulations are known to prolong conidial infectivity and survival and reduce sensitivity to UV radiation compared to aqueous formulations (Moore et al., 1993; Inglis et al., 1995; Alves et al., 1998) their performance has not been consistent under varying environmental conditions and some studies have reported findings similar to results of the present study where aqueous formulations performed same as oil formulations (Ouedraogo et al., 1997). Performance of different species of EPF has also been shown to vary depending on formulation type. For example, Guinossi et al., 2012 reported better dispersal of *B. bassiana* conidia formulated as oil after field application but did not favour better spatial distribution of *M. anisopliae* conidia.

77

One of the limiting factors in commercialization and application of EPF based biopesticides is their slow acting nature and inconsistence performance under environmental conditions (Lacey et al., 2001; St Leger and Wang, 2009; Gašić and Tanović, 2013) and this is being addressed through research on biopesticide formulation. For example Ritu et al., 2012 demonstrated that incorporating bentonite in oil-based formulation of *B. bassiana* improved its efficacy against *Helicoverpa armigera* (Hübner; Lepidoptera: Noctuidae). Addition of white carbon and adjuvants in EPF based biopesticides increases conidia thermotolerance and viability in storage because it absorbs moisture that lowers conidia viability (Kim et al., 2014). Application of Duduthrin did not result in decline in aphid population density as expected with synthetic pesticides and this could be attributed partly due to previously unreported resistance of the aphid to the pesticide since aphids are known to develop resistance to insecticides (Van Emden and Harrington 2007; Simon 2011; Abdallah et al., 2016). Similar observations on the population explosion after use of Duduthrin on aphid species was reported earlier by Bayissa (un published data). In season one there were late aphid infestations (5 weeks after emergence) and low infestation numbers while treatment application commenced at flowering stage (7 weeks after emergence).

In this study application of Duduthrin in both seasons did not prevent cowpea aphid damage while EPF formulations were able to protect the cowpea damage in the second season which was dry and warm. However, the damage was less in the first season because infestation occurred late and plants were old enough and hence less susceptible to damage compared to season 2 where infestation occurred at early stages. Therefore, the stage of crop infestation by aphids determines the level of damage. Evaluation of mycosis caused by oil and aqueous formulations in both seasons showed 95 and 96% and 97 and 96% of the aphid cadavers died as a result of infection by oil and aqueous formulations of the EPF in season 1 and 2 respectively. This reveals the ability of the isolate to cause mortality to *A. craccivora* under field conditions Production of spores on cadavers can trigger secondary infection especially for insects that have aggreggatory behaviour and this can further reduce the population especially under favourable environmental conditions (Pettersson et al., 1998; Purandare and Tenhumberg, 2012). Infection by EPF has been shown to reduce ability of insects to feed and thus cause damage (Roy et al., 2006).

In the present study, the population of natural enemies of aphids remained high in plots treated with EPF in season one and two. Application of the aqueous and oil formulations ICIPE 62 did not negatively affect the population of natural enemies. The ladybird beetles were high in oil and aqueous formulations compared to Duduthrin-treated plots. The aphid predatory spider (*L. decorata*) population was similar in control and both EPF formulations while least in Duduthrin treated plots. The lacewing population was low in Duduthrin treatment in the second season where else the population was high in control and EPF treatments. In previous studies, *M. anisopliae* has been shown to have minimal negative impacts on natural enemies (Zimmermann, 2007). Bidochka and Small (2005) demonstrated that, even though *M. anisopliae* has a wide host range, some strains are specifically pathogenic to certain insect species and not to others. This presents an advantage of the isolate as a potential candidate for commercialization as evaluation of the effect of a potential biopesticide on non-target organisms is one of the criteria used in registration of the biopesticide (Goettel, 1994; Goettel and Hajek, 2001; Montesinos, 2003). Fungal pathogens including EPF play a key role in supplementing the activities of natural enemies in suppressing arthropod pests and therefore positive synergism between them can provide a sustainable option of pest management (Ferguson and Stiling, 1996, Roy and Pell, 2000).

Duduthrin appeared to have negative effects on the natural enemies and this could probably have contributed to higher population of aphids in plots treated with Duduthrin. Previous studies have shown that Lambda-cyhalothrin (active ingredient in Duduthrin) has detrimental effects on *Trichogramma evanescens* (Shoeb, 2010) since the 2 parasites belong to the same order Hymenoptera while other synthetic pyrethroids have been reported to have toxicity on other natural enemies (Tillman and Mulrooney, 2000).

In the first season, Leaf yield data collection commenced 4 weeks after emergence. Treatment application therefore did not influence leaf yield level because by the time aphid population build up to warrant treatment application, yield data collection was advanced. The grain yield was also not influenced by application even though it was lower in Duduthrin treated plots. This could be attributed to the fact that Duduthrin recorded higher numbers of aphids per plant and these could have directly affected the grain yield since treatment applications stopped at leaf senescence while the pods were still developing and since *A. craccivora* is known to attack the pods which remain green and succulent as other plants parts dry, the concentration of aphid colonies on the pods negatively affects their development and subsequent yield. Season 2 was characterized by reduced rainfall, moderate temperatures, early and heavy infestation of *A. craccivora* on cowpea. As a result isolate ICIPE 62 was able to suppress aphid population and enabled the crop to recover from aphid damage and produce more leaf yield compared to synthetic insecticide application. The 2 EPF formulations produced more leaf yields (23.2 and 24.4 kg^{-ha}) respectively compared to untreated control (10.5 and Duduthrin 10.9 kg^{-ha}) respectively. Aphid infestation affects leaf yield quantity and quality because the aphid colonies as well as honeydew produced by aphids and deposited on leaf surfaces renders the leaves unfit for human consumption and also leads to growth of molds that further reduce the harvestable yield. The common practice in urban and peri-urban areas is to uproot cowpea after 1-2 months for leaf consumption and this practice denies the growers the grain yield which earns them more income. This study focused on cowpea as a leafy vegetable but also evaluated grain yield. The study has highlighted the benefit of targeting dual cowpea yield of leaves and grains as opposed to solely focusing on leaf yield. Dual yield makes more economic sense as it earns small-holder farmers more income and the practice of planting and uprooting cowpea while targeting leaf yield should be discouraged.

In conclusion, the *M. anisopliae* isolate ICIPE 62 has the potential to control *A. craccivora* under field conditions when incorporated in integrated pest management strategies as it does not negatively affect beneficial non-target arthropods. However, there is need for further research to develop and optimize formulation types and suitable carriers that eliminates expensive procedures involved in production so to prolong storage period and improve application methods of entomopathogens while at the same time maintaining their efficacy. The application of EPF based biopesticides thus makes economic sense and is recommended as a sustainable management strategy for cowpea aphid. There is also a need to evaluate further the effect of *M. anisopliae* ICIPE 62 on other non-target beneficial arthropods not evaluated in this study fully understand its impact on beneficial organisms.

4.6 References

Abbott W. 1925. A method of computing the effectiveness of an insecticide. Journal of *Journal of Economic. Entomology,* 18 (2):265–267.

Abdallah IS, Abou-Yousef HM, Fouad EA and Kandil MAEH. 2016. The role of detoxifying enzymes in the resistance of the cowpea aphid (*Aphis craccivora* Koch) to thiamethoxam. *Journal of Plant Protection Research,* 56 (1):67-72.

Abukutsa MOO. 2010. African Indigenous Vegetables in Kenya: Strategic Repositioning in the Horticultural Sector. Inaugural Lecture, Jomo Kenyatta University of Agriculture and Technology, Nairobi, Kenya. 30 April.

Alves RT, Bateman RP, Prior C and Leather SR. 1998. Effects of simulated solar radiation on conidial germination of *Metarhizium anisopliae* in different formulations. *Crop Protection,* 17 (8):675-79.

Asante SK, Tamo M and Jackal LEN. 2001. Integrated management of cowpea insect pests using elite cultivars, date of planting and minimum insecticide application. *African Crop Science Journal,* 9 (4):655-665.

Bateman R, Carey M, Moore D and Prior C. 1993. The enhanced infectivity of *Metarhizium flavoviride* in oil formulations to desert locusts at low humidities. *Annals of Applied Biology,* 122:145-152.

Bateman RP and Alves RT. 2000. Delivery systems for mycoinsecticides using oil-based formulations. *Aspects of Applied Biology,* 57:163-170.

Baverstock J, Roy HE, Clark SJ, Alderson PG and Pell JK. 2006. Effect of fungal infection on the reproductive potential of aphids and their progeny. *Journal of Invertebrate Pathology,* 91 (2):136–139.

Bayissa W, Ekesi S, Mohamed SA, Kaaya GP, Wagacha JM, Hanna R, Maniania NK. 2016. Selection of fungal isolates for virulence against three aphid pest species of crucifers and okra. *Journal of Pest Science,* 90:355–68.

Bidochka MJ and Small CS. 2005. Phylogeography of *Metarhizium,* an insect pathogenic fungus. In Vega FE and Blackwell M (Eds.), Insect-fungal associations: Ecology and evolution. New York, NY: Oxford University Press, 28–50 pp.

Bisikwa J, Kawooya R, Ssebuliba JM, Ddungu SP, Biruma M and Okello DK. 2014. Effects of plant density on the performance of local and elite cowpea [*Vigna unguiculata* L. (Walp)] varieties in Eastern Uganda. *African Journal of Agricultural Science and Technology,* 1 (1): 28-41.

Blackman RL and Eastop VF. 2006. Aphids on the world's herbaceous plants and shrubs. Chichester, UK: John Wiley and Sons, 1460 pp.

Chandler D, Bailey AS, Tatchell GM, Davidson G, Greaves J, Grant WP. 2011. The development, regulation and use of biopesticides for integrated pest management. Philosophical transactions of the royal society of London. Series B, *Biological Sciences*, 366 (1573):1987–98.

Ddamulira G, Santos CAF, Obuo P, Alanyo M and Lwanga CK. 2015. Grain yield and protein content of Brazilian cowpea genotypes under diverse Ugandan environments. *American Journal of Plant Sciences*, 6 (13):2074.

Dugje IY, Omoigui LO, Ekeleme F, Kamara AY and Ajeigbe H. 2009. Farmers' guide to cowpea production in West Africa. IITA, Ibadan, Nigeria.

Egho EO and Enujeke EC. 2012. Minimizing insecticide application in the control of insect Pests of Cowpea (*Vigna unguiculata* (L) WALP) in Delta state, Nigeria. *Sustainable Agriculture Research*, 1 (1):87.

Feng MG, Johnson JB and Kish LP. 1990. Survey of entomopathogenic fungi naturally infecting cereal aphids (Homoptera: Aphididae) of irrigated grain crops in southwestern Idaho. *Environmental Entomology*, 19 (5):1534-1542.

Ferguson KI and Stiling P. 1996. Non-additive effects of multiple natural enemies on aphid populations. *Oecologia*, 108:375-379.

Gašić and Tanović B. 2013. Biopesticide formulations, possibility of application and future trends. *Pesticidi i fitomedicina*, 28 (2):97-102.

Goettel MS and Hajek AE. 2000. Evaluation of non-target effects of pathogens used for management of arthropods. In: Wajnberg E, Scott JK, Quimby PC (eds) Evaluating indirect ecological effects of biological control. CABI Publishing, Wallingford, 81–97 pp.

Goettel M. 1994. Host range and specificity in relation to safety of exotic fungi. In 6th International Colloquium on Invertebrate Pathology and Microbial Control, 27th Annual meeting of the society for invertebrate pathology, 325-329 pp.

Guinossi HDM, Moscardi F, Oliveira MCND and Sosa-Gómez DR. 2012. Spatial dispersal of *Metarhizium anisopliae* and *Beauveria bassiana* in soybean fields. *Tropical Plant Pathology*, 37 (1):44-49.

Hajek AE, St. Leger RJ. 1994. Interactions between fungal pathogens and insect hosts. *Annual Review of Entomology*, 39 (1):293–322.

Hall AE. 2012. Phenotyping cowpeas for adaptation to drought. *Frontiers in Physiology*, 3:1-8.

Hornsby AG, Wauchope RD and Herner AE. 1995. Pesticide properties in the environment. Springer, New York, NY, 213 pp.

Inglis GD, Johnson DL and Goettel MS. 1995. Influence of ultraviolet-light protectants on persistence of the entomopathogenic fungus, *Beauveria bassiana*. *Biological Control*, 5 (1):581-590.

Inglis GD, Ivie TJ, Duke GM and Goettel MS. 2000. Influence of rain and conidial formulations on persistence of *Beauveria bassiana* on potato leaves and Colorado potato beetle larvae, *Biological Control*, 18 (1):55-64.

Inglis DG, Jaronski ST, Wraight SP, Beattie A and Watson D. 2002. Use of spray oils with entomopathogens. Spray Oils Beyond 2000: *Sustainable Pest and Disease Management*, 302-312.

Inyang EN, Butt TM, Ibrahim L, Clark, SJ, Pye BJ, Beckett A and Archer S. 1998. The effect of plant growth and topography on the acquisition of conidia of the insect pathogen *Metarhizium anisopliae* by larvae of *Phaedon cochleariae*. *Mycological Research*, 102 (11):1365-1374.

Inyang EN, McCartney HA, Oyejola B, Ibrahim L, Pye BJ, Archer S and Butt TM. 2000. Effect of formulation, application and rain on the persistence of the entomogenous fungus *Metarhizium anisopliae* on oilseed rape. *Mycological Research*, 104 (6):653-661.

International Center of Insect Physiology and Ecology (*icipe*). 2016. ITOC Mbita annual weather data.

Jackai LEN and Daoust RA. 1986. Insect pests of cowpea. *Annual Review of Entomology* (31): 95-119.

Kenya Meteorological Department. 2017. Annul weather data. http://www.meteo.go.ke/ /index.php?q=datarequest._Copyright Â© 2015 Kenya Meteorological Department.

Kim JS, Lee SJ and Lee HB. 2014. Enhancing the thermotolerance of entomopathogenic *Isaria fumosorosea* SFP-198 conidial powder by controlling the moisture content using drying and adjuvants. *Mycobiology*, 42(1):59-65.

Kithure JGN, Murung JI, Tum PK, Wanjau RN and Thoruwa CL. 2017. Fate of Lambda-Cyhalothrin in Kales, Tomatoes and Cabbage from Rural setting in Kenya. *International Journal of Scientific Research and Innovative Technology*, (4) 2:2313-3759.

Lacey LA, Grzywacz D, Shapiro-Ilan DI, Frutos R, Brownbridge M and Goettel MS. 2015. Insect pathogens as biological control agents: back to the future. *Journal of Invertebrate Pathology*, 132: 1-41.

Maniania NK. 1993. Evaluation of three formulations of *Beauveria bassiana* (Bals.) Vuill. for control of the stem borer *Chilo partellus* (Swinhoe) (Lep., Pyralidae). *Journal of Applied Entomology*, 115 (1-5):266-272.

Milner RJ. 1997. Prospects for biopesticides for aphid control. *BioControl*, 42 (1):227-239.

Montesinos E. 2003. Development, registration and commercialization of microbial pesticides for plant protection. *International Microbiology*, 6 (4):245-252.

Moore D, Bridge PD, Higgins PM, Bateman RP and Prior C. 1993. Ultra-violet radiation damage to *Metarhizium flavoviride* conidia and the protection given by vegetable and mineral oils and chemical sunscreens. *Annals of Applied Biology*, 122 (3):605-616.

Moore D, Bateman RP, Carey M and Prior C. 1995. Long-term storage of *Metarhizium flavoviride* conidia in oil formulations for the control of locusts and grasshoppers. *Biocontrol Science and Technology*, 5 (2):193-200.

Mucheru-Muna M, Pypers P, Mugendi D, Kung'u J, Mugwe J, Merckx R and Vanlauwe B. 2010. A staggered maize–legume intercrop arrangement robustly increases crop yields and economic returns in the highlands of Central Kenya. *Field Crops Research*, 115 (2):132-139.

Mweke A, Ulrichs C, Nana P, Akuste KS, Fiaboe KKM, Maniania NK and Ekesi S. 2018. Evaluation of the Entomopathogenic Fungi *Metarhizium anisopliae*, *Beauveria bassiana* and *Isaria* sp. for the Management of *Aphis craccivora* (Hemiptera: Aphididdae). *Journal of Economic Entomology*, 111 (4): 1587–1594.

Mweke A, Ulrichs C, Maniania KN, Ekesi S. 2016. Integration of entomopathogenic fungi as biopesticide for the management of cowpea aphid (*Aphis craccivora* Koch). *African Journal of Horticultural Science*, 9:14–31.

Nussenbaum AL, Lewylle MA, Lecuona RE. 2013. Germination, radial growth and virulence to boll weevil of entomopathogenic fungi at different temperatures. *World Applied Sciences Journal*, 25:1134–1140.

Ofuya TI. 1997. Control of the bean aphid *Aphis craccivora* Koch (Homoptera: Aphididae) in cowpea, *Vigna unguiculata* (L.) Walp. *Integrated Pest Management Reviews* 2 (4):199-207.

Omoigui LO, Ekeuro GC, Kamara AY, Bello LL, Timko MP and Ogunwolu GO. 2017. New sources of aphids (*Aphis craccivora* (Koch)) resistance in cowpea germplasm using phenotypic and molecular marker approaches. *Euphytica*, 213 (8): 178.

Ouedraogo A, Fargues J, Goettel MS and Lomer CJ. 1997. Effect of temperature on vegetative growth among isolates of *Metarhizium anisopliae* and *M. flavoviride. Mycopathologia*, 137 (1):37-43.

Pettersson J, Karunaratne S, Ahmed E and Kumar V. 1998. The cowpea aphid, *Aphis craccivora*, host plant odours and pheromones. *Entomologia Experimentalis et Applicata*, 88: 177-184.

Purandare SR and Tenhumberg B. 2012. Influence of aphid honeydew on the foraging behaviour of *Hippodamia convergens* larvae. *Ecological Entomology*, 37: 184–192.

R Core Team. 2013. R: A language and environment for statistical computing. R Foundation for Statistical Computing, Vienna, Austria. URL. http://www.R-project.org/.

Ritu A, Anjali C, Nidhi T, Sheetal P and Deepak B. 2012. Biopesticidal formulation of *Beauveria bassiana* effective against larvae of *Helicoverpa armigera. Journal of Biofertilizers and Biopesticides*, 3(3).

Roy HE and Pell JK. 2000. Interactions between entomopathogenic fungi and other natural enemies: implications for biological control. *Biocontrol Science and Technology*, 10 (6):737-752.

Roy HE, Steinkraus DC, Eilenberg J, Hajek AE and Pell JK. 2006. Bizarre interactions and endgames: entomopathogenic fungi and their arthropod hosts. *Annual. Review of Entomology*, 51: 331-357.

Rusike JG, van den Brand S, Boahen K, Dashiell S, Kantengwa J, Ongoma DM, Mongane G, Kasongo ZB, Jamagani R, Aidoo R and Abaidoo R. 2013. Value chain analyses of grain legumes in N2Africa: Kenya, Rwanda, Eastern DRC, Ghana, Nigeria, Mozambique, Malawi and Zimbabwe (No. 1.2. 6, 1.3. 4). N2Africa.

Saidi M, Itulya FM, Aguyoa JN, Mshenga PM, Owour G. 2010. Effects of cowpea leaf harvesting initiation time on yields and profitability of a dual-purpose sole cowpea and cowpea-maize intercrop. *Electronic Journal of Environmental, Agricultural and Food Chemistry*, 9 (6):1134-1144.

Santi LE, Silva LA, da Silva WOB, Corrêa APF, Rangel DEN, Carlini CR, Schrank A and Vainstein MH. 2011. Virulence of the entomopathogenic fungus *Metarhizium anisopliae* using soybean oil formulation for control of the cotton stainer bug, *Dysdercus peruvianus. World Journal of Microbiology and Biotechnology*, 27 (10):2297-2303.

Schreiner I. 2000. Cowpea Aphids, Agricultural pests of the pacific. ADAP 2000-6, Reissued February 2000. ISBN 1-931435-09-X.

Schumacher V and Poehling HM. 2012. In vitro effect of pesticides on the germination, vegetative growth, and conidial production of two strains of *Metarhizium anisopliae. Fungal Biology*, 116 (1):121-132.

Selvaraj K Kaushik HD, Gulati R and Sharma SS. 2010. Bioefficacy of *Beauveria bassiana* (Balsamo) Vuillemin against *Hyadaphis coriandri* (Das) on coriander and *Aphis craccivora* Koch on fenugreek. *Journal of Biological Control*, 24 (2):142-146.

Shah PA, Pell JK. 2003. Entomopathogenic fungi as biological control agents. *Applied Microbiology and Biotechnology*, 61 (5-6):413–23.

Shapiro SS and Wilk MB.1965. An analysis of variance test for normality. *Biometrika*, 53 (2):591–611.

Shoeb MA. 2010. Effect of Some insecticides on the immature stages of the egg parasitoid *Trichogramma evanescens*. *Egyptian Academic Journal of Biological Sciences*, 3:31-38.

Simon JYu. 2011. The toxicology and biochemistry of insecticides. CRC Press, Taylor & Francis Group, Boca Raton, USA, 211 pp.

Souleymane A, Aken'Ova ME, Fatokun CA and Alabi OY. 2013. Screening for resistance to cowpea aphid (*Aphis craccivora* Koch) in wild and cultivated cowpea (*Vigna unguiculata* Walp.) accessions. *International Journal of Science, Environment and Technology*, 2 (4):611-621.

Tillman PG and Mulrooney JE. 2000. Effect of selected insecticides on the natural enemies *Coleomegilla maculata* and *Hippodamia convergens* (Coleoptera: Coccinellidae), *Geocoris punctipes* (Hemiptera: Lygaeidae), and *Bracon mellitor, Cardiochiles nigriceps*, and *Cotesia marginiventris* (Hymenoptera: Braconidae) in cotton. *Journal of Economic Entomology*, 93 (6):1638-1643.

Trehan I, Benzoni NS, Wang AZ, Bollinger LB, Ngoma TN, Chimimba UK, Stephenson KB, Agapova SE, Maleta KM and Manary MJ. 2015. Common beans and cowpeas as complementary foods to reduce environmental enteric dysfunction and stunting in Malawian children: study protocol for two randomized controlled trials. *Trials*, 16 (1):520.

Van Emden H, Harrington R. 2007. Aphids as Crop Pests. CABI North American Office, Cambridge, USA, 699 pp.

Waddington SR, Li X, Dixon J, Hyman G and De Vicente MC. 2010. Getting the focus right: production constraints for six major food crops in Asian and African farming systems. *Food Security*, 2 (1):27-48.

Wraight SP and Ramos ME. 2002. Application parameters affecting field efficacy of *Beauveria bassiana* foliar treatments against Colorado potato beetle *Leptinotarsa decemlineata*. *Biological Control*, 23 (2):164-178.

Wraight SP, Filotas MJ and Sanderson JP. 2016. Comparative efficacy of oil-oil, wettable-powder, and unformulated-powder preparations of *Beauveria bassiana* against the melon aphid *Aphis gossypii*, *Biocontrol Science and Technology*, 26 (7):894-914.

Zimmermann G. 2007. Review on safety of the entomopathogenic fungus *Metarhizium anisopliae*. *Biocontrol Science and Technology*, 17 (9):879–920.

5.0 INTEGRATED MANAGEMENT OF *APHIS CRACCIVORA* IN COWPEA USING INTERCROPPING AND APPLICATION OF ENTOMOPATHOGENIC FUNGI UNDER FIELD CONDITIONS

Submitted in Annals of Entomology as: Integrated Management of *Aphis craccivora* in cowpea using intercropping and application of entomopathogenic fungi under field conditions

Mweke A, Ulrichs C, Akutse KS, Nana P, Maniania NK, Fiaboe KKM, Ekesi S

5.1 Abstract

Cowpea aphid (*Aphis craccivora* Koch) is a major pest of cowpea in the tropics. Intercropping cowpea and cereals has been used in management of *A. craccivora* with minimal success. Combining intercropping and application of pesticides in management of the pest has been shown to be effective but the chemicals are associated with health risks and environmental pollution. Use of fungal based biopesticides is an attractive option where cowpea is grown mainly as a vegetable because biopesticides are not toxic to the environment and beneficial organisms. This study evaluated the effect of combining application of entomopathogenic fungi (EPF) *Metarhizium anisopliae* isolate ICIPE 62 in cowpea-maize intercrop in management of *A. craccivora* under field conditions. There were six treatment as follows: cowpea monocrop treated with EPF, cowpea-maize intercrop treated with EPF, untreated cowpea monocrop, untreated cowpea-maize intercrop, cowpea monocrop treated with Duduthrin (Lambda-cyhalothrin (1.75 g AI)) and cowpea-maize intercrop treated with Duduthrin. The treatments were replicated 4 times and the experiment was carried out for three seasons. The fungus was applied at the rate (1×10^{12} conidia ml^{-1}) in oil formulation. The first season was wet with heavy and frequent rainfall and cooler temperatures, the treatments did not reduce aphid population and cowpea crop damage did not differ significantly. The treatments also did not impact the leaf and grain yield. The second season was relatively drier with higher temperatures and the cowpea-maize intercrop treated with EPF recorded the lowest infestation and the least cowpea damage and the leaf yield was comparable to cowpea monocrop treated with EPF despite lower plant population in the intercrop. In the third season which recorded higher rainfall than season 2 but lower than season 1, the cowpea-maize intercrop treated with EPF recorded lowest infestation and damage. Leaf yield was similar to season two. Application of Duduthrin either in monocrop or intercrop did not reduce aphid infestation neither did it protect the crop against damage. This study has demonstrated the potential of integrated management of *A. craccivora* through application of EPF in a cowpea-maize intercrop under field conditions confer yield benefit to farmers.

Key words: Cowpea aphid, *Metarhizium anisopliae*, yield, Duduthrin, ICIPE 62

5.2 Introduction

Cowpea (*Vigna unguiculata* L. Walp) is primarily a tropical crop that originated in Africa and has spread to other parts of the world (Afiukwa et al., 2013). The crop is mostly grown as an intercrop with cereals and farmers are able to harvest even when cereals fail due to inadequate rainfall because it is drought tolerant (Dahmardeh et al., 2010; Boukar et al., 2011; Hassan, 2013). The crop is an important leafy vegetable and a valuable source of affordable proteins, vitamins and income to rural households (Oyewale, 2013; Trehan et al., 2015). In Kenya cowpea is one of the most important indigenous vegetable in production and consumption (Abukutsa-Onyango, 2010; HCD, 2016). The area under production of these indigenous vegetables has been increasing (Cernansky, 2015).

Cowpea aphid (*Aphis craccivora* Koch) is a polyphagous pest of cowpea that attacks the crop, feeding on all plant parts and leading to yield losses (Blackman and Eastop, 2000; Obopile and Ositile, 2010; Keatinge et al., 2015). Cowpea aphid feeding damage includes sucking and removal of plant sap that reduces amounts of nutrients and water available to the crop and transmission of plant viruses. Aphids feeding induces symptoms that include chlorosis, stunting, and delayed onset of flowering and even plant death when infestations are high especially at seedling stage (Blackman and Eastop, 2000; Obopile and Ositile, 2010). Among the management strategies of *A. craccivora* the use of chemical insecticides is ranked first by farmers because there are many chemical insecticides registered for use in management of aphids including *A. craccivora* (Egho, 2010). Even though aphids are susceptible to pesticides, application of the chemicals does not always result in effective suppression of their population because of their high fecundity and have been reported to develop resistance to some of the chemicals (Soliman, 2015; Abdallah et al., 2016; Mokbel et al., 2017). Besides synthetic chemicals pose health risks due to toxic residues especially on leafy vegetables which are harvested in short intervals (Keatinge et al., 2015; Mweke et al., 2016) and also kill natural enemies of *A. craccivora* leading to pest resurgence and need for more pesticides application (Jackai and Daoust, 1986; Abdallah et al., 2016). Synthetic pyrethroids including Cypermethrin, Aphacypermethrin Deltamethrin and Lambda-cyhalothrin are the most commonly used pesticides. Lambda-cyhalothrin based insecticides are used in management of aphids in Kenya and come under different trade names and are easy to access and use. Intercropping cowpea with cereals like maize, sorghum and millet been used as a strategy in management of cowpea insect pests including *A. craccivora* though does not completely control the pest (Karungi et al., 2000; Nabirye et al., 2003; Hassan, 2013). Therefore, enhancement of this strategy by monitored application of insecticides has been shown to offer benefits to farmers especially where cowpea is grown as a grain legume (Afun et al., 1991; Egho, 2012). However, where cowpea is grown as a leafy vegetable application of insecticides increases the risk of consuming pesticide residues since the leaves are harvested regularly and hence there is need to adopt safer pest management strategies.

86

Different groups of entomopathogenic fungi (EPF) have known pathogenicity to different insect pests and their use as biological control agents can reduce reliance on chemical insecticides (Roy et al., 2010; St Leger and Wang, 2010; Khan et al., 2016, Valero-Jiménez et al., 2016; Zhang et al., 2017).Use of biopesticides based on EPF in management of *A. craccivora* in vegetables is a good alternative to synthetic insecticides because the pest is susceptible to different groups of entomopathogenic fungi (Ekesi et al., 2000; Mweke et al., 2018). A number of EPF based biopesticide products for management of aphids are available in Europe and Americas but are few in Africa and none has been registered for use on *A. craccivora* (Faria and Wraight, 2007; Mweke et al., 2016). Several species of *Metarhizium* have been identified as being pathogenic to *A. craccivora* both in laboratory and in field conditions (Sahayaraj and Borgio 2010; Saranya et al., 2010; Mweke et al., 2018). Biopesticides derived from EPF have advantages of being compatible with integrated pest management (IPM) strategies (Lopes et al., 2011) though their slow activity under field conditions slows down their wide spread use (St Leger and Wang, 2009; Gašić and Tanović, 2013). Performance of biopesticides based on EPF can be enhanced by using them in combination with other pest control strategies like cultural control.

Though EPF based biopesticides have been recommended for use in IPM, no previous studies have been carried out to assess their efficacy when combined with intercropping cowpea and maize in control of *A. craccivora*. This study therefore evaluated the efficacy of combining EPF application and intercropping cowpea and maize in management of *A. craccivora* under field conditions.

5.3 Materials and methods

5.3.1 Fungal culture and inoculum preparation
The fungus *Metarhizium anisopliae* isolate ICIPE 62 with a known pathogenicity to *A. craccivora* and obtained from *icipe*'s Arthropod Germplasm Centre was used in this study. Long grain rice in Milner bags (60 x 35 cm) sterilized by autoclaving for 1 hour at 120°C were used as substrate for production of fungal conidia. The autoclaved substrate was cooled to room temperature in plastic buckets 35 (Ø) × 25 (width) ×15 cm (depth) before inoculating with a 3-day-old culture of blastopores (50 ml) after which it was covered with sterile polyethylene bags. The inoculated substrate culture was incubated for 21 days at ambient conditions (20-26°C, 40-70% RH) (Maniania, 1993). After the incubation period the bag was removed then dried at room temperature for 5 days. Conidia were harvested by sifting the substrate through a sieve (295-μm mesh size) and stored in a refrigerator (4-6°C, 40-50% RH) for before being used in field experiments. The fungus viability was evaluated before field treatment application by spread plating 100 μl of conidial at a concentration of 3×10^6 conidia ml^{-1} in Sabouraud dextrose agar (SDA) plates and then incubating them at 26 ± 2°C in darkness for 18 h after which percent fungal spore germination was determined by counting randomly 100 selected conidia on a cover slip under a light microscope (400x) (Goettel and Inglis, 1997). The conidia germ tubes that were

at least as long as twice the diameter of the conidium were scored as viable or germinated (Schumacher and Poehling, 2012). Conidial germination >90% after 18 h on SDA and was considered adequate for use in the field trials. Conidia concentration in 1 gram was determined by dissolving 0.1 gram of conidia in 10ml of sterile Triton water then serial diluting to × 100 after which the mixture was vortexed and 1 ml pipetted into a hemocytometer and spores counted under the microscope. The amount of conidia in grams required to produce a concentration of 1×10^{12} was determined from spores in 0.1ml.

5.3.2 Lambda-cyhalothrin Duduthrin® l.75EC

The pesticide used in this study was acquired from local agro-chemical outlets. Before treatment application, 65ml of the pesticide was mixed with 20 litres of clean water in a knapsack sprayer and 0.05 % (1ml) Integra (sticker, Greenlife Crop Protection Africa Ltd) added and mixture thoroughly mixed before application.

5.3.3 Experimental sites

The experiment was carried out at *icipe*'s Mbita point campus, Homabay County, Western Kenya for two seasons. In the first season the experiment was carried out between March and June 2016 in a field located at 00.42931S, 034.20604 E, 1140 meters above sea level (m.s.l). The mean annual rainfall between March and June was 131 mm while the minimum and maximum temperatures were 20°C and 25.2 °C respectively with relative humidity ranging between 60 and 70%. In the second season the experiments were done between May and August 2016 and 1155 m.s.l where the mean rainfall was 53.7mm. Minimum and maximum temperatures were 23.7 °C and 29.5 °C and relative humidity ranged between 60 and 65%. The third season was carried experiment was carried out between July and October 2016. During this season, the mean rainfall was 80.5mm and the minimum and maximum temperatures were 25.8°C and 29.3°C while relative humidity was between 65 and 70% (http://www.meteo.go.ke/).

5.3.4 Crops

The land was prepared by ploughing and harrowing before planting. Ex-Luanda, a local land race which is susceptible to *A. craccivora* obtained from *icipe*'s germplasm collection was used in the experiment. Maize variety PHB 3253 (Pioneer Hi-Bred Kenya Limited) was used in this study and was acquired from local agro-vet shops. Cowpea was planted in alternate rows with maize in plots measuring 10m x 10m; spacing for cowpea was 20-cm intra-row by 75-cm inter-row with two seed sown per hill. The cowpea plants were later thinned to one plant per hill at 14 days after emergence giving a plant population of 66,666 plants per hectare in monocrop and 33,333 plants per hectare in intercrop. Spacing for maize was 30-cm intra-row and 90-cm inter rows and plant population per hectare was 18,518 plants per hectare in the intercrop. Overhead irrigation (sprinkler) was used for the first 3 weeks to support the crop as it was relatively dry in January during planting and irrigation discontinued after the onset of the rains. Weeding was done twice a month before the crop established and smothered the weeds and the frequency of weeding reduced to once a month. The crop was then left for natural aphid infestation.

5.3.5 Treatments, layout and design
The experiment was carried out for 3 seasons and each season 1 experimental field was planted. The experiment had six treatments as follows: (i) cowpea monocrop, untreated, (ii) cowpea - maize intercrop, untreated, (iii) cowpea monocrop treated with the fungus (ICIPE 62), (iv) cowpea-maize intercrop, treated with the fungus (ICIPE 62), (v) cowpea monocrop, treated with Duduthrin and (vi) cowpea-maize intercrop, treated with Duduthrin. Oil formulation of the fungus was used and the spores were suspended in water containing 0.05% Integra (sticker, Greenlife Crop Protection Africa Ltd) with 0.1% nutrient agar, 0.1% glycerin and 0.5% molasses added as protectants and attractant respectively (Maniania, 1993). The insecticide Duduthrin® 1.75EC (Twiga Chemical Industries Ltd) was applied at the rate of 1.75g (Active Ingredient (AI)) ha^{-1} with 0.1% Integra. Control treatment were sprayed with water containing 0.05% Integra, 0.1% nutrient agar, 0.05% molasses and 0.1% glycerin without any fungal conidia and any insecticide solution. Conidia were applied at the rate of 1 x 10^{12} conidia ml^{-1}. In season 1the spray applications started on day 56 after planting due to late aphids' infestation and thereafter done on a weekly basis for a period of 6 weeks. Second and third seasons, treatment application began 21 days after planting because aphid infestation occurred early compared to season 1. The fungus formulations and the insecticide were applied with different knapsack sprayers with target output of 350 L ha^{-1} and spraying was done late in the evening between 17:00 and 18:00 h. The experimental design was randomized block design with four replications.

5.3.6 Evaluation of treatments
Two leaflets, each from the base and the top from 20 randomly selected cowpea plants in the middle rows were sampled from each plot for aphid infestation assessment. The aphids were dislodged from the host plant with a fine hair brush into a vial containing 70% alcohol, labelled and thereafter counted in the laboratory. Sampling was done on weekly basis from day 7 after planting until the cowpea leaves begun to dry out.

5.3.7 Assessment of cowpea leaf damage
Leaf damage (leaf quality and fitness for human consumption) were visually assessed using the following scale (Benchasri, 2009)
0 = visual damage on leaves and flower buds < 10%; 1 = visual damage on leaves and flower buds 10-25%; 2 = visual damage on leaves and flower buds 26-50%; 3 = visual damage on leaves and flower buds 51-75%; 4 = visual damage on leaves and flower buds 76-100%

5.3.8 Aphid mortality Assessment

Mortality of aphids induced by the fungus was assessed by picking 30 aphids from each EPF treated plot and transferring them into plastic dishes (11.3 cm (Ø) × 4 cm (depth) lined with moist filter paper and placing sterilized cowpea leaves in the dishes to serve as food for the aphids. Muslin cloth with apertures (300 x 300 μ) was placed around the container mouth before placing the cover to allow for free air circulation. The dishes were kept at room temperature and mortality observed daily for 1 week. The leaves serving as food for the aphid were removed and replaced with fresh ones daily. Dead aphids were collected and placed in petri-dishes with sterilized moist filter papers and kept at room temperature before observation under the dissecting microscope for mycosis.

5.3.9 Cowpea leaf vegetable and grain yield assessment

Cowpea leaf vegetable yield data was collected at 7 weekly starting day 21 after planting (Saidi et al., 2010a). The total leaf vegetable weight for each treatment was calculated by pooling together the fresh leaf weights obtained per for very treatment at the different leaf harvesting dates, and expressed in kilograms per hectare (kg ha^{-1}). Dry cowpea grain yield was obtained by picking mature pods, sun drying and threshing and recording the grain weight using electronic weighing balance and yield computed and also expressed in kg ha^{-1}.

5.3.10 Assessment of natural enemies of *A. craccivora*

Ladybird beetles, spiders, lacewing and parasitoids were the natural enemies of *A. craccivora* encountered in this study. Apart from parasitoid the other natural enemies were assessed by counting their numbers on randomly selected cowpea plants in each plot. Parasitoids were assessed by collecting 20 mummies per plot and transferring them to perforated petri-dishes and keeping them at room temperature and the number of parasitoids emerging from the mummies recorded.

5.4 Statistical analysis

The aphid infestation density and natural enemies count data, aphid damage assessment score data, leaf and grain yield data were first log transformed before subjecting the data to ANOVA and means separated using Tukey HSD. Aphid mortality data induced by the EPF were corrected for natural mortality (Abbott, 1925), tested for normality test (Shapiro and Wilk, 1965) and arcsine transformed before subjecting the data to ANOVA and means separated using Tukey HSD. Data were analyzed using R software (R Core, 2013).

5.5 Results

5.5.1 Effect of intercropping maize and cowpea and treatment application on aphid infestation

In the first season 6 weeks of treatment application did not result in significant difference in aphid infestation (number of aphids per plant) among the treatments (F=1.57, df=5, P=0.18) (Table 5.1). In the second season the aphid infestation varied significantly between the treatments (F=7.2, df=5, P=0.001) (Table 5. 1). Cowpea maize intercrop treated with EPF recorded the lowest aphid density per plant followed by cowpea monocrop treated with EPF, Untreated cowpea-maize intercrop and untreated cowpea monocrop while cowpea monocrop treated with Duduthrin and cowpea-maize intercrop recorded the highest aphid infestation. A similar trend was observed in the third season where the treatment application resulted in significant differences in aphid infestation levels among the treatments (F=8.26, df=5, P=0.001). Cowpea-maize intercrop treated with EPF recorded least aphid infestation per plant while there was no significant difference between the other treatments (Table 5. 1).

Table 5.1: Mean *Aphis craccivora* population density per plant after treatment for 3 seasons

Treatments	Aphid density per plant
Season 1(wet)	
Cowpea monocrop Control	11.8 ± 1.6a
Cowpea monocrop Duduthrin	9. 2 ± 1.5a
Cowpea monocrop EPF	8.1 ± 2.83a
Cowpea-Maize Intercrop Control	11.3 ± 1.4a
Cowpea-Maize Intercrop Duduthrin	9.8 ± 1.5a
Cowpea-Maize Intercrop EPF	8.1 ± 2.3a
Season 2 (dry)	
Cowpea monocrop Control	7.8 ± 3.6bc
Cowpea monocrop Duduthrin	15.3 ± 5.8a
Cowpea monocrop EPF	5.3 ± 2.4bc
Cowpea-Maize Intercrop Control	6.3 ± 3bc
Cowpea-Maize Intercrop Duduthrin	10.4 ± 3.4b
Cowpea-Maize Intercrop EPF	2.7 ± 1c
Season 3 (off season)	
Cowpea monocrop Control	13.9 ± 3.4a
Cowpea monocrop Duduthrin	9.7 ± 2.8ab
Cowpea monocrop EPF	8.1 ± 4bc
Cowpea-Maize Intercrop Control	12.9 ± 5.9a
Cowpea-Maize Intercrop Duduthrin	9.5 ± 3.3ab
Cowpea-Maize Intercrop EPF	5.4 ± 2c

Means followed by the same letter within a column are not significantly different by Tukey HSD in each season.

5.5.2 Aphid mortality and mycosis induced by EPF application in the field

Mortality and infection of aphid by EPF was evaluated for all the six treatments. In the first season, mortality induced by EPF on the monocrop and the intercrop was 80.4 ± 3.1 and 79.2 ± 3 while percent mycosis of dead aphids was 75.17 ± 3.8 and 72.64 ± 4 respectively and both mortality (F=0.08, df=1, P= 0.05) and mycosis (F=0.25, df=1, P=0.04) differed significantly between the treatments (Table 5.2).

Table 5.2: Mean mortality and mycosis mycosis of *Aphis craccivora* induced by the EPF

Treatment	Mortality (%)	Mycosis (%)
Season 1(wet)		
Cowpea monocrop Control	0.0 ± 0.0b	0.0 ± 0.0b
Cowpea monocrop Duduthrin	0.0 ± 0.0b	0.0 ± 0.0b
Cowpea monocrop EPF	80.4 ± 3.1a	75.17 ± 3.8a
Cowpea-Maize Intercrop Control	0.0 ± 0.0b	0.0 ± 0.0b
Cowpea-Maize Intercrop Duduthrin	0.0 ± 0.0b	0.0 ± 0.0b
Cowpea-Maize Intercrop EPF	79.2 ± 3a	72.64 ± 4a
Season 2 (dry)		
Cowpea monocrop Control	0.0 ± 0.0b	0.0 ± 0.0b
Cowpea monocrop Duduthrin	0.0 ± 0.0b	0.0 ± 0.0b
Cowpea monocrop EPF	83.29 ± 2.18a	79.16 ± 1.6a
Cowpea-Maize Intercrop Control	0.0 ± 0.0b	0.0 ± 0.0b
Cowpea-Maize Intercrop Duduthrin	0.00 ± 0.0b	0.0 ± 0.0b
Cowpea-Maize Intercrop EPF	84.41 ± 1.8a	81.12 ± 1.2a
Season 3 (off season)		
Cowpea monocrop Control	0.0 ± 0.0b	0.0 ± 0.0b
Cowpea monocrop Duduthrin	0.0 ± 0.0b	0.0 ± 0.0b
Cowpea monocrop EPF	88.1 ± 2.2a	82.25 ± 2.15a
Cowpea-Maize Intercrop Control	0.0 ± 0.0b	0.0 ± 0.0b
Cowpea-Maize Intercrop Duduthrin	0.0 ± 0.0b	0.0 ± 0.0b
Cowpea-Maize Intercrop EPF	88.46 ± 2.3a	84.41 ± 1.9a

Means followed by the same letter within a column are not significantly different by Tukey HSD in each season.

In the second season both the mortality and mycosis differed significantly between the treatments (F=24.4, df=1, P=0.001) and mycosis (F=0.15, df=1, P=0.05) (Table 5.2). In the third season the mortality (F=279.68, df=5, P=0.001) and mycosis (F=199.71, df=5, P=0.001) differed significantly among the treatments (Table 5. 2). There was no mycosis detected in untreated control and Duduthrin treated plots implying there was no cross contamination between the treatments. There was no mycosis detected in untreated control and Duduthrin treated plots implying there was no cross contamination between the treatments.

5.5.3 Aphid damage on Cowpea

In season 1 the treatments were not able to protect the cowpea crop from damage by *A. craccivora* and there was no significant difference in damage between the treatments (F=0.75, df=5, P=0.59) (Table 5.3).

Table 5.3: Cowpea plant damage by *Aphis craccivora* for 3 seasons

Treatment	Damage (%)
Season 1(dry)	
Cowpea monocrop Control	$32.0 \pm 10.0a$
Cowpea monocrop Duduthrin	$36.4 \pm 12.3a$
Cowpea monocrop EPF	$25.2 \pm 10.6a$
Cowpea-Maize Intercrop Control	$31.4 \pm 12.0a$
Cowpea-Maize Intercrop Duduthrin	$31.1 \pm 10.2a$
Cowpea-Maize Intercrop EPF	$30.1 \pm 7.9a$
Season 2 (wet)	
Cowpea monocrop Control	$23.4 \pm 5.6b$
Cowpea monocrop Duduthrin	$23.6 \pm 6.8b$
Cowpea monocrop EPF	$17.7 \pm 3.5b$
Cowpea-Maize Intercrop Control	$20.1 \pm 4.0b$
Cowpea-Maize Intercrop Duduthrin	$44.5 \pm 8.9a$
Cowpea-Maize Intercrop EPF	$8.2 \pm 0.2c$
Season 3 (off season)	
Cowpea monocrop Control	$33.5 \pm 9.3\ a$
Cowpea monocrop Duduthrin	$24.9 \pm 6.7ab$
Cowpea monocrop EPF	$21.7 \pm 3.5ab$
Cowpea-Maize Intercrop Control	$33.9 \pm 12.7a$
Cowpea-Maize Intercrop Duduthrin	$26.5 \pm 8.9ab$
Cowpea-Maize Intercrop EPF	$13.3 \pm 0.1c$

Means followed by the same letter within a column are not significantly different by Tukey HSD for each season.

In the second season, the least damage on cowpea was observed in cowpea-maize intercrop treated with EPF 8.2% while the highest damage was recorded in cowpea-maize intercrop (44.5%) treated with Duduthrin and the damage was significantly different among the treatments (F=6.42, df-5, P=0.001) (Table 5. 3). The damage caused by aphids differed significantly in the third season (F=4.11, df=5, P=0.001) with the lowest damage observed in cowpea maize intercrop treated with EPF whereas highest damage was recorded in untreated cowpea-maize intercrop followed by untreated cowpea monocrop (Table 5.3).

5.5.4 Natural enemies associated with *Aphis craccivora*

The natural enemies of *A. craccivora* that were observed in the 3 seasons included the ladybird beetles, Spiders (*Leucocage decorata*), lacewing and Parasitoid (*Aphidius colemani*). In the first season the ladybird beetles were significantly different among the treatments (F=3.4, df=5, P=0.04) with the highest numbers recorded in untreated cowpea monocrop and untreated cowpea-maize intercrop. In the second season, the number of the ladybird beetles significantly between the treatments (F=13.9, df=5. P=0.001). The untreated cowpea-maize intercrop least number of ladybird beetles. Treatment application in the third season resulted in significant differences in the number of the ladybird beetles among the treatments (F= 4.48, df=3, P=0.001) with the highest population observed in untreated cowpea monocrop (Table 5.4).

There was no significant difference between the treatments in the number of *L. decorata* the first (F=1.34, df=5, P=0.25) and the second seasons (F= 1.23, df=5, P=0.3) (Table 3). Third season recorded significant difference in the number of *L. decorata* between the treatments (F=2.45, df=5, P=0.03) (Table 5.4).

The lacewing numbers recorded in season 1 were not significantly different between the treatments (F=1, df=5, P=0.42), but in season 2 there was significant difference in their numbers among the 6 treatments (F=2.6, df=5, P=0.02). The application of the different treatments in the third season did not produce significant differences in the number of lacewing (F= 1.22, df=5, P=0.3) (Table 5.4).

The aphid parasitoid (*A. colemani*) numbers were not significantly different between the treatment in the three seasons (F=0.5, df=5, P=0.78, F=0.11, df=5, P=0.1 and F=1.48, df=5, P=0.19) for season 1, 2 and 3 respectively (Table 5.4).

Table 5.4: Beneficial arthropods population density for 3 seasons

Treatments	Lady bird beetles	Spiders (*Leucocage decorata*)	Lacewing	Parasitoid (*Aphidius colemani*)
Season 1(wet)				
Cowpea monocrop Control	2.7 ± 0.4a	1.6 ± 0.3b	2.6 ± 0.6a	0.5 ± 0.2a
Cowpea monocrop Duduthrin	1.3 ± 0.2b	2.3 ± 0.8a	1.8 ± 0.5b	0.3 ± 0.2a
Cowpea monocrop EPF	1.5 ± 0.2ab	2.4 ± 0.5a	1.7 ± 0.3b	0.5 ± 0.2a
Cowpea-Maize Intercrop Control	2.3 ± 1.0ab	2.7 ± 0.5a	1.2 ± 0.3b	0.4 ± 0.2a
Cowpea-Maize Intercrop Duduthrin	1.3 ± 0.5b	1.3 ± 0.3b	1.4 ± 0.4b	0.2 ± 0.1a
Cowpea-Maize Intercrop EPF	1.6 ± 0.3ab	2.8 ± 0.6a	1.7 ± 0.4b	0.5 ± 0.1a
Season 2(wet)				
Cowpea Control	10.6 ± 2.2a	1.9 ± 0.5a	0.4 ± 0.2ab	0.21 ± 0.1a
Cowpea Duduthrin	10.1 ± 1.7a	1.2 ± 0.4a	0.5 ± 0.2ab	0.21 ± 0.1a
Cowpea EPF	6.5 ± 1.1a	0.8 ± 0.3a	0.3 ± 0.2b	0.17 ± 0.07a
Cowpea-Maize Intercrop Control	2.6 ± 0.8b	1.4 ± 0.3a	0.5 ± 0.2ab	0.13 ± 0.04a
Cowpea-Maize Intercrop Duduthrin	10.4 ± 1.5a	2.2 ± 0.5a	1 ± 0.4a	0.17 ± 0.05a
Cowpea-Maize Intercrop EPF	6.5 ± 1.1a	1.2 ± 0.2a	1.2 ± 0.2a	0.13 ± 0.1a
Season 3(off season)				
Cowpea Control	3.9 ± 0.7a	1 ± 0.2a	0.6 ± 0.2a	0.02 ± 0.01a
Cowpea Duduthrin	2.9 ± 0.1ab	0.7 ± 0.2ab	0.3 ± 0.1a	0.07 ± 0.01a
Cowpea EPF	2 ± 0.5b	0.3 ± 0.1b	0.2 ± 0.1a	0.32 ± 0.02a
Cowpea-Maize Intercrop Control	2 ± 0.5b	0.5 ± 0.1ab	0.1 ± 0.06a	0.00 ± 0.0a
Cowpea-Maize Intercrop Duduthrin	2.5 ± 0.6b	0.7 ± 0.2ab	0.3 ± 0.2a	0.09 ± 0.01a
Cowpea-Maize Intercrop EPF	2.4 = 0.5b	0.4 ± 0.1ab	0.2 ± 0.1a	0.14 ± 0.05a

Means followed by the same letter within a column are not significantly different by Tukey HSD in each season.

5.5.6 Cowpea leaf and grain yield

The green leaf yield in the first season significantly different between the treatments (F=4.58, df=5, P=0.001) (Figure 5.1). Untreated cowpea monocrop, cowpea, cowpea monocrop treated with Duduthrin and EPF produced the highest yield while cowpea-maize intercrop treated with EPF recorded lower yield.

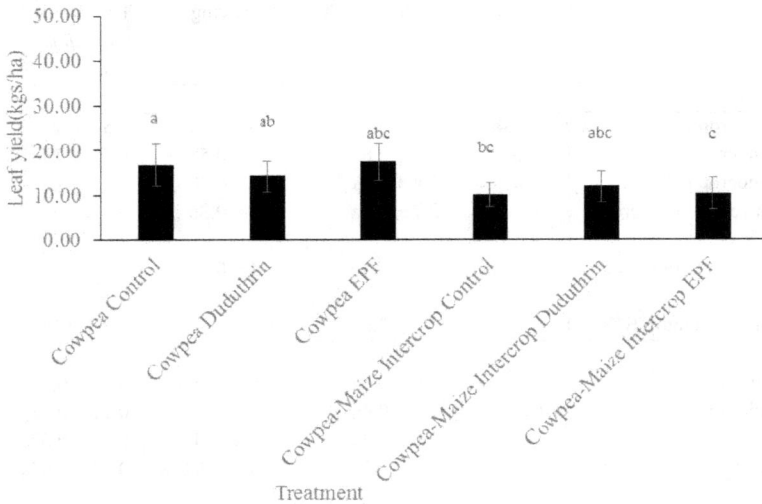

Figure 5.1: Season 1(wet) leaf yield in kgs ha^{-1}. Bars represent means ± SE at p<0.05

In the second season among the cowpea monocrops, the cowpea monocrop treated with EPF recorded higher leaf yield while in the intercrop cluster, the cowpea-maize intercrop recorded higher green leaf yield and there was significant difference between the treatments (F=6.3, df=5, P=0.001) (Figure 5. 2).

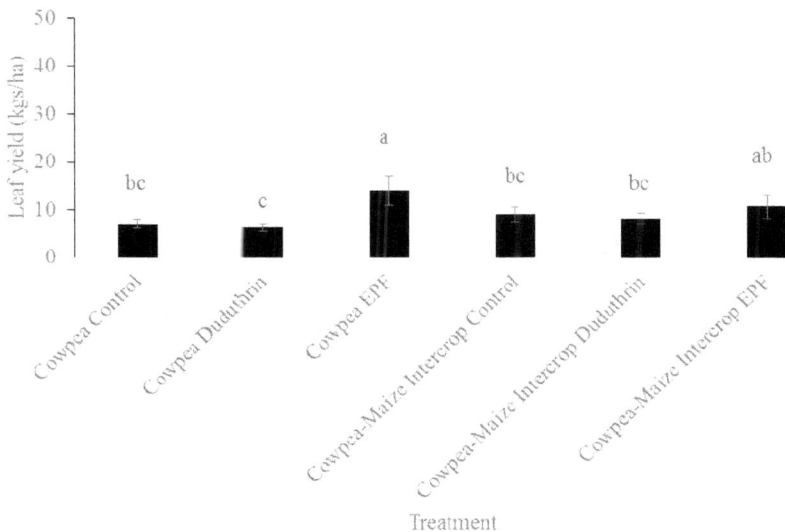

Figure 5.2: Season 2 (dry) leaf yield in kgs ha^{-1}. Bars represent means ± SE at p<0.05

In the third season the leaf yield differed significantly between the six treatments (F=2, df=5, P=0.08), (Figure 5.3). The cowpea monocrop treated with EPF recorded highest yield followed by cowpea-maize treated with EPF while cowpea monocrop treated with Duduthrin produced the lowest yield.

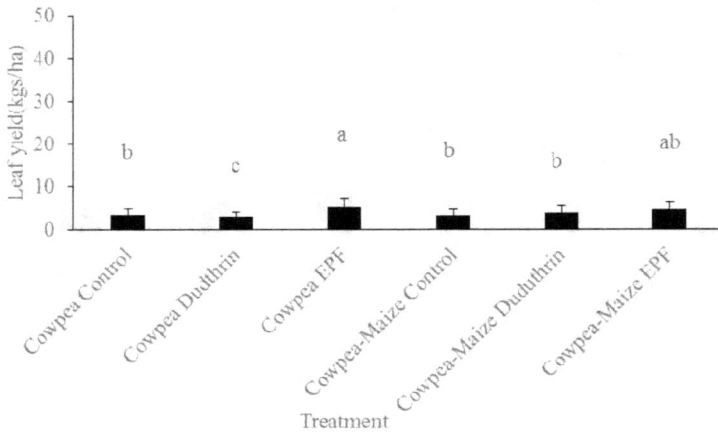

Figure 5. 3: Season 3 (off season) leaf yield in kgs ha^{-1}. Bars represent means ± SE at p<0.05

In the first season the cowpea grain yield was significantly different among the 6 treatments (F=8, df=5, P=0.001). Treatment application did not influence the leaf yield in season one because the yield was similar among the monocrops and also similar among the intercrops (Figure 5.4). All the monocrops with the highest cowpea plant population produced same yield while the intercrops with same but lower plant population compared to monocrops also recorded similar leaf yield.

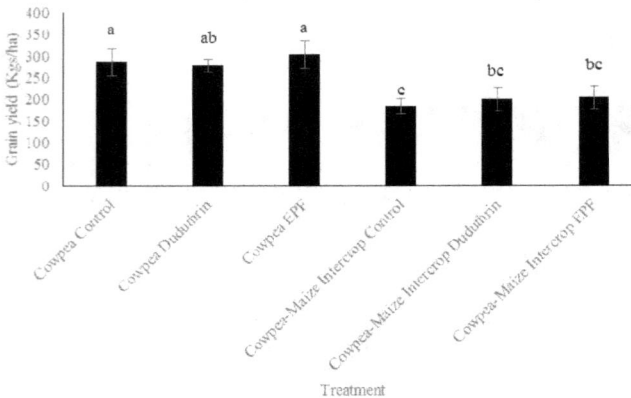

Figure 5.4: Season 1 (wet) cowpea grain yield in kgs ha^{-1}. Bars represent means ± SE at p<0.05.

In season two however, the cowpea monocrop treated with EPF recorded higher yield and there was significant difference observed in the treatments. (F=29.3, df=5, P=0.001) (Figure 5.5). Untreated cowpea-maize intercrop, cowpea-maize intercrop treated with Duduthrin recorded the lowest leaf yield. The yield in cowpea monocrop treated with Duduthrin was comparable to the untreated cowpea-maize intercrop and cowpea-maize intercrop treated with Duduthrin despite the former having higher cowpea plant population.

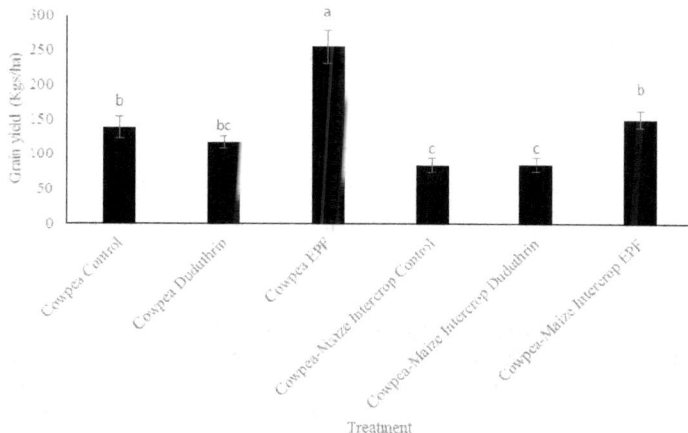

Figure 5.5: Season 2 (dry) cowpea grain yield in kgs ha^{-1}. Bars represent means ± SE at p<0.05

In the third season the cowpea monocrop treated with EPF recorded highest overall grain yield. Cowpea-maize intercrop treated with EPF produced the highest grain yield among the intercrops and was comparable to cowpea monocrop treated with Duduthrin even though the former had higher plant population. The yield was significantly different among the treatments (F=8.3, df=5, P=0.001) (Figure 5.6).

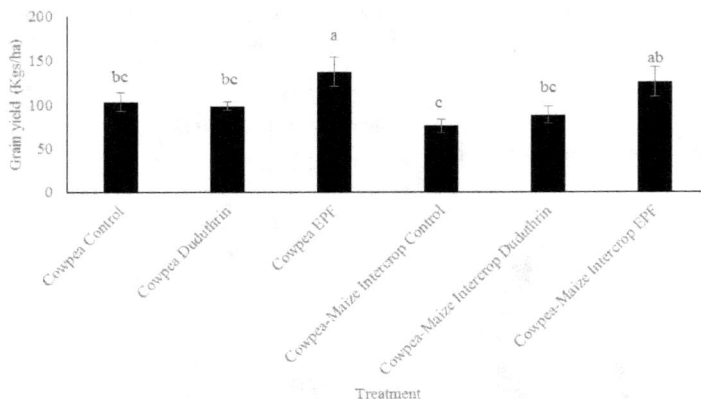

Figure 5.6: Season 3 (off season) cowpea grain yield in kgs ha⁻¹. Bars represent means ± SE at p<0.05

5.6 Discussion

In this study, first season recorded heavy and frequent rainfall and late infestation of aphids. All the treatments did not reduce aphid infestation though untreated cowpea monocrop recorded slightly higher aphid density. In the second season, there was reduced rainfall and early and heavy aphid infestation, the treatments effects on the aphid population was noticeable and the treatment combination of cowpea-maize intercrop was more effective in reducing aphid infestations per plant and recorded the least aphid density after six weeks of treatment. Untreated cowpea-maize intercrop and cowpea-maize intercrop treated with Duduthrin did not reduce aphid infestations effectively compared to cowpea-maize intercrop treated with EPF, in fact cowpea-maize intercrop- Duduthrin treatment recorded higher levels of infestation than untreated cowpea-maize intercrop while the cowpea monocrop treated with Duduthrin also had higher aphid densities than untreated cowpea monocrop. The third season recorded lower rainfall than first season but was higher than second season, but the temperatures were similar to those in season 2. In this season the cowpea-maize intercrop treated with EPF isolate ICIPE 62 recorded least aphid population density per plant and was comparable to cowpea-maize intercrop treated with EPF. The performance of untreated cowpea-maize intercrop, cowpea monocrop treated with Duduthrin, untreated cowpea-maize intercrop and cowpea-maize intercrop treated with Duduthrin was similar and did not reduce aphid infestation effectively. The good performance of the combination of intercropping maize and cowpea and application of EPF can be attributed to several factors. Intercropping results in increased relative humidity and reduced light penetration into the lower canopy crop (Kyamanywa and Ampofo, 1988; Terao, 1997). Performance of EPF based biopesticides is known to be affected by relative humidity and UV light (Ferron et al., 1991; Hajek and St. Leger, 1994; Jaronski, 2010). Therefore, improved relative humidity and reduced

100

fungal spore degradation due to light interception in the intercrop could have enhanced performance of the EPF resulting in suppression of *A. craccivora* population. Similar findings were reported by Ekesi et al., 1999 who demonstrated better control of legume flower thrips (*Megalurothrips sjostedti* (Trybom; Thysanoptera: Thripidae) in cowpea intercropped with maize and treated with EPF (*M. anisopliae*).

Even though intercropping strategy in insect pest management including aphids has been widely researched (Ogenga-latigo et al., 1992; Hassan, 2013, Kisetu et al., 2014), studies including the present one have demonstrated that use of this strategy alone does not guarantee successful pest management. In the present study intercropping cowpea and maize without application of EPF and application of Duduthrin (a synthetic pyrethroid) in cowpea intercropped with maize did not reduce aphid infestations in all the three seasons. Duduthrin (Lambda-cyhalothrin) which is registered for aphid management did not control the aphids either in the monocrop or intercrop. Previous studies have reported similar findings where Duduthrin application did not result in lower *A. craccivora* infestations or control (Mweke et al. unpublished data, Bayissa et al. personal communication). This could be due to development of resistance by *A. craccivora* since aphids are known to develop resistance to chemicals after years of exposure (Van Emden and Harrington, 2007; Simon, 2011; Abdallah et al., 2016). Therefore, research on improvement of the use of intercropping strategy has been developed and has focused on combination of several strategies like intercropping and monitored application of pesticides (Afun et al., 1991, Egho and Enujeke; 2012). However, deleterious effects of chemical pesticides have encouraged search for safer alternatives. Application of EPF based biopesticides in cowpea-maize intercrop systems can provide alternatives to use of synthetic chemical pesticides and could provide a more sustainable management of aphids since the biopesticides are user friendly, safe to environment and do not harm natural enemies of aphids. The *M. anisopliae* isolate used in this study (ICIPE 62) has known pathogenicity to several aphid species including *A. craccivora* (Bayissa et al., 2016a; Mweke et al., 2018) and has been developed by *icipe* and commercialized by Real IPM for use in management of aphids (http://www.realipm.com/).

During the first season, the treatments had no effect on the damage of cowpea by *A. craccivora* as there was no difference between the treatments, however, untreated cowpea monocrop recorded higher damage compared to other treatments. In the second season of this study where aphid infestation occurred early after crop emergence and the weather was relatively dry with lower rainfall and elevated temperatures, the damage on the cowpea crop was higher in cowpea monocrop treated with Duduthrin though the damage was not significantly different from cowpea monocrop treated with EPF, untreated cowpea monocrop and untreated cowpea-maize intercrop while cowpea-maize intercrop treated with EPF recorded the least damage. In the third season, all the treatments recorded similar damage except cowpea-maize intercrop treated with EPF which provided adequate protection against *A. craccivora*. Intercropping cowpea and cereals like maize and sorghum has several advantages including increased yield, improvement in soil fertility, and better utilization of resources as well as insect pest management (Ofori and Stern, 1987; Saxena et al., 1989; Hassan, 2013). In intercropped systems pests are visually disturbed and tend to stay for shorter times on the hosts due to disruptive effect of landing on non-host plants and also reduces their survival and damage (Root, 1973; Vandermeer, 1989). Intercropping cowpea and cereals creates physical barrier against aphids (Jackai et al., 1985). In this study intercropping alone without treatment application or intercropping and application of Duduthrin did not protect the crop against damage by aphids. The cowpea landrace used in this study has a known susceptibility to *A. craccivora* and damage is bound to occur even at low levels of infestation (Omoigui et al., 2017). Aphid population builds up fast within a short period of time in favourable environmental conditions due to their reproductive nature (parthenogenesis and viviparity) and this increases damage in susceptible plants (Omoigui et al., 2017). The inability of the intercrop to reduce aphid infestation could also be attributed to the cropping practice used in this study. Cowpea and maize were planted simultaneously and hence the maize could not offer the physical barrier or the visual disturbance effects (Vandermeer, 1989; Jackai et al., 1985). Even though it has been shown that staggered planting is able to protect cowpea form insect pests in an intercrop system (Afun, 1991; Kisetu et al., 2014), the applicability of this cropping system is not feasible in arid and drier areas where cowpea is a major crop since early and simultaneous cropping of multiple crops is the norm. This enables the crops to benefit from the little available moisture.

During the three seasons the natural enemies of *A. craccivora* encountered were ladybird beetles, spiders (*Leucocage decorata*), lacewing and parasitoid (*Aphidius colemani*) and their numbers varied across the treatments. In the first season, the highest number of ladybird beetles was recorded in untreated cowpea monocrop and the least was cowpea monocrop treated with Duduthrin. Second season recorded the highest number of ladybirds among all the three seasons. This could be attributed to the fact that aphid infestation occurred early in the second season and the population build up was higher implying that the aphid population could sustain higher number of the predatory ladybird beetles, while in the third season the highest number of the ladybird beetles were in untreated cowpea monocrop plots. Across the three seasons, spider population was higher in season one compared to season two and three and although the untreated cowpea monocrop recorded lower population in season one, the observation was different in season two and three. The lacewing population was highest in season one but lower in season two and three. In season one the untreated cowpea monocrop recorded higher population of lacewing among the treatments.

The parasitoid *A. colemani* was the lowest in population among the natural enemies and did not appear to be affected by the treatments. The fact that the parasitoid numbers were not significant across the treatments for 3 seasons implies that, the treatments did not influence their numbers and rather there were other factors in play. The parasitoid is affected by different biotic and biotic factors in the natural environment. Temperature affects development of the parasitoid (Colinet et al., 2007). Relative humidity also affects fecundity, hatching and longevity (Prado et al., 2015).

Presence of other natural enemies like predators which were plenty in the study site increases competition for aphid and reduces available prey for oviposition. Additionally, some predators consume parasitized aphids thus reducing the parasitoid population over time (Bilu and Coll, 2007) and the parasitoid does not discriminate between predator invested or predator free plants (Brodeur and Rosenheim, 2000). *Aphidius colemani* is highly sensitive to residual pesticides. The lower number of the parasitoid could be attributed to the sensitivity of the genera to pesticides (Shipp et al., 2000; Araya et al., 2010) and since the experimental site has been used for different experiments for a long time some of which involved use of synthetic pesticides, this could have reduced their population gradually. The authors therefore hypothesize that a combination of either of these factors could have led to low parasitoid numbers in the experimental site

In the present study, both cowpea leaf and grain yield was evaluated for three seasons. In the first season the leaf yield was higher in monocrops compared to intercrops irrespective of treatments. The higher leaf yield in monocrops was attributed to higher cowpea plant population in monocrops. The lack of significant difference in leaf yield in the first season in monocrops and intercrops was attributed to the late and low aphid infestation where the infestations occurred after leaf yield data collection was almost complete therefore the treatments did not affect the yield. Similarly, the cowpea damage in the first season was not influenced by the treatments and this also explains why the leaf yield was not significant between the treatments.

In the second season, aphid infestations occurred early after crop emergence (seven days after emergence (DAE)) and treatment applications commenced fourteen DAE, the cowpea monocrop treated with EPF recorded higher leaf yield among the monocrops and overall. This implies that application of EPF (*M. anisopliae* isolate ICIPE 62) was able to protect the crop despite early heavy infestation by *A. craccivora* and was able to produce higher leaf yield. This offers an advantage for use of the EPF based biopesticide since it able to control *A. craccivora* and confer leaf yield benefit to the farmers. Biopesticides are not associated with toxic residues in food products and are suitable for use in vegetables which are harvested frequently. Duduthrin, a synthetic pyrethroid registered for field control of sucking pests including aphids did not confer yield advantage as expected. This was explained by the fact the application of Duduthrin did not result in decline in aphid infestation (aphids per plant) and did not protect cowpea plants form aphid damage. Aphid damage on leaf yield is not only through direct feeding on the leaves but also affects quality due to production of honeydew on the leaves that also reduces marketable leaf yield. In the third season, aphid infestation occurred twenty-one DAE and treatment commenced immediately thereafter. The season was characterized by intermittent rainfall. The cowpea monocrop treated with EFF produced higher leaf yield. This could be attributed to the ability of ICIPE 62 to suppress aphid population and that resulted in better yields.

The cowpea grain yield in the first season was not influenced by treatments both in the mono and intercrops and the authors hypothesize that this could have been due to late infestation and heavy rainfall that meant that the damage was minimal since aphid are easily washed off plant surfaces by rainfall or overhead irrigation water. In the second season cowpea monocrop treatment produced more grain yield while among the intercrops, treatment combination with EPF produced more grain yield than the untreated control as well as Duduthrin. In the third season cowpea-maize intercrop treated with EPF produced higher grain yield while maize intercrop treated with EPF and Duduthrin produced similar yields. *Aphis craccivora* attacks cowpea in all stages of its development from seedling to podding stage and is one of the key limiting factors in cowpea production (Blackman and Eastop 2006; Souleymane et al., 2013). Application of EPF *M. anisopliae* ICIPE 62 in cowpea maize intercrop was able to suppress the *A. craccivora* population compared to Duduthrin (Lambda-cyhalothrin), a synthetic insecticide popularly used in management of aphid. Although performance of EPF based biopesticides is limited by its slow acting nature and environmental factors (Leng et al., 2011; Roy et al., 2006; Eyheraguibel et al., 2010; Jaronski, 2010), the current study has demonstrated that integrating EPF based biopesticide in an intercrop system coupled with favourable weather conditions can offer effective *A. craccivora* control and confer yield benefits to farmers. The *M. anisopliae* isolate (ICIPE 62) has been demonstrated to be pathogenic to several aphid species under laboratory and screenhouse conditions (Bayissa et al., 2016a, Mweke et al., 2018). This study has therefore demonstrated the potential of the isolate to control *A. craccivora* under field conditions in an integrated management system. The isolate also did not negatively affect the natural enemies of *A. craccivora* and this presents another advantage for the commercialization of the isolate. The non-target effect observed in this study has been reported earlier under laboratory conditions (Bayissa et al., 2016b). Cowpea is usually grown for dual purpose i.e. leafy vegetable and grain (Saidi et al., 2010b) and this presents an advantage because farmers can derive dual benefits from the yields. Though EPF based biopesticides are slow acting and do not produce immediate control and their efficacy is dependent on environmental conditions (Roy et al., 2006; Eyheraguibel et al., 2010, Jaronski, 2010), they can be used in an integrated management systems.

In conclusion, this study has demonstrated the efficacy of combining intercropping and application of EPF as an alternative approach that can be used in management of aphids in vegetables. Use of synthetic pesticides in vegetables is disadvantageous because of the need to adhere to post harvest intervals (PHI). Vegetables are harvested frequently and use of the synthetic pesticides leads to yield loss during observation of PHIs and this also increases food safety risks besides the environmental pollution and killing of non-target beneficial arthropods by the pesticides.

5.7 References

Abbott WS. 1925. Method of computing the effectiveness of an insecticide. *Journal of Economic Entomology*, 18 (2):265–267.

Abdallah IS, Abou-Yousef H.M, Fouad EA and Kandil MAEH. 2016. The role of detoxifying enzymes in the resistance of the cowpea aphid (*Aphis craccivora* Koch) to thiamethoxam. *Journal of Plant Protection Research*, 56 (1):67-72.

Abukutsa MOO. 2010. African indigenous vegetables in Kenya: Strategic repositioning in the horticultural sector. Inaugural lecture, Jomo Kenyatta University of Agriculture and Technology, Nairobi, Kenya.

Afiukwa CA, Ubi BE, Kunert KJ, Titus EJ and Akusu JO. 2013. Seed protein content variation in cowpea genotypes. *World Journal of Agricultural Sciences*, 1 (3):094-099.

Afun JVK, Jackai LEN and Hodgson CJ. 1991. Calendar and monitored insecticide application for the control of cowpea pests. *Crop Protection*. 10 (5):363-370.

Araya JE, Araya M and Guerrero MA. 2010. Effects of some insecticides applied in sublethal concentrations on the survival and longevity of *Aphidius ervi* (Haliday) (Hymenoptera: Aphidiidae) Adults. *Chilean Journal of Agricultural Research*, 70 (2):221-227.

Baverstock J, Roy HE, Clark SJ, Alderson PG and Pell JK. 2006. Effect of fungal infection on the reproductive potential of aphids and their progeny. *Journal of Invertebrate Pathology*, 91 (2):136–139.

Bayissa W, Ekesi S, Mohamed SA, Kaaya GP, Wagacha JM, Hanna R and Maniania NK. 2016a. Selection of fungal isolates for virulence against three aphid pest species of crucifers and okra. *Journal of Pest Science*, 90:355-68.

Bayissa W, Ekesi S, Mohamed SA, Kaaya GP, Wagacha JM, Hanna R and Maniania NK. 2016b. Interactions among vegetable-infesting aphids, the fungal pathogen *Metarhizium anisopliae* (Ascomycota: Hypocreales) and the predatory coccinellid *Cheilomenes lunata* (Coleoptera: Coccinellidae). *Biocontrol Science and Technology*, 26:274-290.

Bilu E and Coll M. 2007. The importance of intraguild interactions to the combined effect of a parasitoid and a predator on aphid population suppression. *BioControl*, 52:753–763.

Blackman RL and Eastop VF. 2006. Aphids on the world's herbaceous plants and shrubs. Chichester, UK: John Wiley & Sons, 1460 pp.

Brodeur J and Rosenheim JA. 2000. Intraguild interactions in aphid parasitoids. *Entomologia Experimentalis et Applicata*, 97(1):93-108.

Boukar O, Massawe F, Muranaka S, Franco J, Maziya-Dixon B, Singh B and Fatokun C. 2011. Evaluation of cowpea germplasm lines for protein and mineral concentrations in grains. *Plant Genetic Resources*, 9(4):515-522.

Cernansky R. 2015. Super vegetables. *Nature*. 522 (7555):146.

Colinet H, Boivin G and Hance T. 2007. Manipulation of parasitoid size using the temperature-size rule: Fitness consequences. *Oecologia*, 152:425–433.

Dahmardeh M, Ghanbari A, Syahsar BA and Ramrodi M. 2010. The role of intercropping maize (*Zea mays* L.) and Cowpea (*Vigna unguiculata* L.) on yield and soil chemical properties. *African Journal of Agricultural Research*, 5 (8):631-636.

Egho EO and Enujeke EC. 2012. Minimizing insecticide application in the control of insect pests of cowpea (*Vigna unguiculata* (L) WALP) in Delta State, Nigeria. *Sustainable Agriculture Research*, 1 (1):87.

Ekesi S, Maniania NK, Ampong-Nyarko K and Onu I. 1999. Effect of intercropping cowpea with maize on the performance of *Metarhizium anisopliae* against *Megalurothrips sjostedti* (Thysanoptera: Thripidae) and predators. *Environmental Entomology*, 28 (2):1154-1161.

Ekesi S, Akpa AD, Onu I, Ogunlan MO. 2000. Entomopathogenicity of *Beauveria bassiana* and *Metarhizium anisopliae* to the cowpea aphid, *Aphis craccivora. Archives of Phytopathology and Plant Protection*, 33 (2):171–180.

Eyheraguibel B, Richard C, Ledoigt G and Ter Halle A. 2010. Photoprotection by plant extracts: a new ecological means to reduce pesticide photodegradation. *Journal of Agriculture and Food Chemistry*, 58 (17):9692-9696.

Faria MRD and Wraight SP. 2007. Mycoinsecticides and mycoacaricides: a comprehensive list with worldwide coverage and international classification of formulation types. *Biological Control*, 43 (3):237-256.

Ferron P, Fargues J and Riba G. 1991. Fungi as microbial insecticides against pests. Handbook of applied mycology, 2, 665-706 pp.

Galidevara S, Reineke A and Koduru UD. 2016. In vivo expression of genes in the entomopathogenic fungus *Beauveria bassiana* during infection of lepidopteran larvae. *Journal of Invertebrate Pathology*, 136:32-34.

Gašić and Tanović B. 2013. Biopesticide formulations, possibility of application and future trends. *Pesticidi i fitomedicina*, 28(2):97-102.

Goettel MS and Inglis GD. 1997. Fungi: hyphomycetes. In: Lacey LA (ed) manual of techniques in insect pathology, 213-249 pp.

Hajek AE, St. Leger RJ. 1994. Interactions between fungal pathogens and insect hosts. *Annual Review of Entomology*, 39 (1):293–322.

Hassan S. 2013. Effect of variety and intercropping on two major cowpea (*Vigna unguiculata* L. Walp) field pests in Mubi, Adamawa state, Nigeria. *International Journal of Agricultural Research and Development*, 1 (5):108-109.

Horticultural Crops Directorate (HCD). 2016. Annual report, 2015-2016.

Humber RA. 1991. Fungal pathogens of aphids. In: Peters D.C, Webster J.A, Chlouber C.S (eds). Aphid-plant interactions: Populations to molecules, Agricultural Experiment Station, Division of Agriculture, Oklahoma State University, Stillwater, 45–56 pp.

Jackai LEN, Singh, SR, Raheja AK and Wiedijk F. 1985. Recent trends in the control of cowpea pests in Africa. In: S.R. Singh and K.O. Rachie (eds) Cowpea Research, Production and Utilization. Chichester: John Wiley & Sons, 234-43 pp.

Jackai LEN and Daoust RA. 1986. Insect pests of cowpeas. *Annual Review of Entomology,* 31 (1):95-119.

Jaronski, ST. 2010. Ecological factors in the inundative use of fungal entomopathogens. *BioControl*, 55 (1):159-185.

Jandricic SE, Mattson NS, Wraight SP and Sanderson JP. 2014. Within-Plant distribution of *Aulacorthum solani* (Hemiptera: Aphididae), on various greenhouse plants with implications for control. *Journal of Economic Entomology*, 107 (2):697-707.

Karungi J, Adipala E, Kyamanywa S. Ogenga-Latigo M, Oyobo N, Jackai L. 2000. Pest management in cowpea. Part 2. Integrating planting time, plant density and insecticide application for management of cowpea field insect pest in Eastern Uganda. *Crop Protection*, 19 (4) 237-245.

Keating JDH, Wang JF, Dinssa FF, Ebert AW, Hughes JDA, Stoilova T, Nenguwo N, Dhillon NPS, Easdown WJ, Mavlyanova R and Tenkouano. 2015. Indigenous vegetables worldwide: their importance and future development. *Acta Horticulturae*, 1102:1-20.

Kenya Meteorological Department. 2017. Annul weather data.
http://www.meteo.go.ke//index.php?q=datarequest. Copyright Â© 2015 Kenya Meteorological Department.

Khan S, Nadir S, Lihua G, Xu J, Holmes KA and Dewen Q. 2016. Identification and characterization of an insect toxin protein, Bb70p, from the entomopathogenic fungus, *Beauveria bassiana*, using *Galleria mellonella* as a model system. *Journal of Invertebrate Pathology*, 133: 87-94.

Kisetu E, Nyasasi BT and Nyika M. 2014. Effect of cropping systems on infestation and severity of field pests of cowpea in Morogoro, Tanzania. *Modern Journal of Agriculture*, 1(1):1-9.

Kyamanywa S and Ampofo JKO. 1988. Effect of cowpea/maize mixed cropping on the incident light at the cowpea canopy and flower thrips (Thysanoptera: Thripidae) population density. *Crop Protection*, 7 (3):186-189.

Lopes RB, Pauli G, Mascarin GM and Faria M. 2011. Protection of entomopathogenic conidia against chemical fungicides afforded by an oil-based formulation. *Biocontrol Science and Technology*, 21 (2):125-137.

Maniania NK. 1993. Evaluation of three formulations of *Beauveria bassiana* (Bals.) Vuill. for control of the stem borer *Chilo partellus* (Swinhoe) (Lep. Pyralidae). *Journal of Applied Entomology*, 115 (1-5):266-272.

Mokbel ESMS, Swelam ESH, Radwan EMM and Kandil MAE. 2017. Role of metabolic enzymes in resistance to chlorpyrifos-methyl in the cowpea aphid, *Aphis craccivora* (Koch). *Journal of Plant Protection Research*, 57(3):288-293.

Mweke A, Ulrichs C, Maniania K.N and Ekesi S. 2016. Integration of entomopathogenic fungi as biopesticide for the management of cowpea aphid (*Aphis craccivora* Koch). *African Journal of Horticultural Science*. 9:14–31.

Mweke A, Ekesi S, Maniania NK, Fiaboe KKM, Ulrichs C. 2017. Performance of *Metarhizium anisopliae* (Metsch.) Sorok and *Beauveria bassiana* (Bals.) Vuill. isolates against cowpea aphid (*Aphis craccivora* Koch) in cowpea under field conditions, Tropentag Conference, 20th -23rd September, Bonn, Germany.

Mweke A, Ulrichs C, Nana P, Akutse KS, Fiaboe KKM, Maniania NK and Ekesi S. 2018. Evaluation of the entomopathogenic fungi *Metarhizium anisopliae, Beauveria bassiana* and *Isaria* sp. for the management of *Aphis craccivora. Journal of Economic Entomology*, 111 (4): 2018, 1587–1594.

Nabirye J, Nampala P, Ogenga-Latigo MW, Kyamanywa S, Wilson H, Odeke V, Iceduna C. and Adipala E. 2003. Farmer-participatory evaluation of cowpea integrated pest management (IPM) technologies in Eastern Uganda. *Crop Protection*, 22 (1):31-38.

Obopile M and Ositile B. 2010. Life table and population parameters of cowpea aphid, *Aphis craccivora* Koch (Homoptera: Aphididae) on five cowpea *Vigna unguiculata* (L. Walp.) varieties. *Journal of Pest Science*, 83: 9-14.

Ofori F and Stern WR. 1987. Cereal–legume intercropping systems. In Advances in agronomy. Academic Press. Vol. 41, 41-90 pp.

Ogenga-Latigo MW, Ampofo JKO and Balidawa CW. 1992. Influence of maize row spacing on infestation and damage of intercropped beans by bean aphid (*Aphis fabae* Scop.) I. Incidence of Aphids. *Field Crop Research*, 30 (1-2):111-121.

Omoigui LO, Ekeuro GC, Kamara AY, Bello LL, Timko MP and Ogunwolu GO. 2017. New sources of aphids (*Aphis craccivora* (Koch) resistance in cowpea germplasm using phenotypic and molecular marker approaches. *Euphytica*, 213: 178.

Oyewale RO and Bamaiyi LJ. 2013. Management of cowpea insect pests. *Scholars Academic Journal of Biosciences*, 1 (5):217-226.

Prado SG, Jandricic SE and Frank SD. 2015. Ecological interactions affecting the efficacy of *Aphidius colemani* in greenhouse crops. *Insects*, 6(2):538-575.

R Core 2013. Team: A language and environment for statistical computing. R Foundation for Statistical Computing, Vienna, Austria. URL http://www.R-project.org/.

Root RB. 1973. Organization of plant-arthropod association in simple and diverse habitats: the fauna of collards (*Brassica oleracea*). *Ecological Monographs*, 43(1):95-1.

Roy HE, Steinkraus DC, Eilenberg J, Hajek AE and Pell JK. 2006. Bizarre interactions and endgames: entomopathogenic fungi and their arthropod hosts. *Annual Review of Entomology*, 51:331-357.

Roy HE, Brodie EL, Chandler D, Goettel MS, Pell JK, Wajnberg E and Vega F. 2010. Deep space and hidden depths: understanding the evolution and ecology of fungal entomopathogens. *Biological control*, 55:1–6.

Saidi M, Itulya FM, Aguyoh JN and Mshenga PM. 2010a. Yields and profitability of a dual-purpose sole cowpea and cowpea-maize intercrop as influenced by cowpea leaf harvesting frequency. *Journal of Agricultural and Biological Science*, 5 (5): 65-71.

Saidi M, Itulya FM, Aguyoh JN, Mshenga PM, Owour G. 2010b. Effects of cowpea leaf harvesting initiation time on yields and profitability of a dual-purpose sole cowpea and cowpea-maize intercrop. *Electronic Journal of Environmental, Agricultural and Food Chemistry*, 9(6): 1134-1144.

Saxena KN, Okay AP, Reddy KS, Omolo EO and Ngode L. 1989. Insect pest management and socio-economic circumstances of small-scale farmers for food crop production in western Kenya: A case study. *International Journal of Tropical Insect Science*, 10 (4):443-462.

Schumacher V and Poehling HM. 2012. In vitro effect of pesticides on the germination, growth, and conidial production of two strains of *Metarhizium anisopliae*. *Fungal Biology*, 116(1): 121-132.

Shapiro SS and Wilk MB.1965. An analysis of variance test for normality. *Biometrika*, 53 (2):591–611.

Shipp JL, Wang K and Ferguson G. 2000. Residual toxicity of avermectin b1 and pyridaben to eight commercially produced beneficial arthropod species used for control of greenhouse pests. *Biological Control*, 17:125–131.

Soliman EAFM. 2015. Biochemical and molecular bases of resistance mechanism in cowpea aphid *Aphis craccivora* (Koch) to Pirimicarb and Thiamethoxam. Doctoral dissertation, Cairo, University, Egypt.

Souleymane A, Aken'Ova ME, Fatokun CA and Alabi OY. 2013. Screening for resistance to cowpea aphid (*Aphis craccivora* Koch) in wild and cultivated cowpea (*Vigna unguiculata* Walp.) accessions. *International Journal of Science, Environment and Technology*, 2 (4): 611-621.

St. Leger RJ and Wang CS 2009. Entomopathogenic fungi and the genomics era. In: Stock S.P, Vandenberg J, Glazer I, Boemare N (eds). Insect pathogens: molecular approaches and techniques. CABI, Oxfordshire, 365–400 pp.

St Leger R.J and Wang C. 2010. Genetic engineering of fungal biocontrol agents to achieve greater efficacy against insect pests. *Applied Microbiology and Biotechnology*, 85 (4):901–7.

Terao T, Watanabe I, Matsunaga R, Hakoyama S and Singh BB. 1997. Agro-physiological constraints in intercropped cowpea: an analysis: In B. B. Singh, D. R. Mohan Raj, K. E. Dashiell, and L.E.N. Jackai [eds.], Advance in cowpea research. Sayce, Devon, UK, 129-140 pp.

Valero-Jiménez CA, Wiegers H, Zwaan BJ, Koenraadt CJ and van Kan JA. 2016. Genes involved in virulence of the entomopathogenic fungus *Beauveria bassiana*. *Journal of Invertebrate Pathology*, 133: 41-49.

Van Emden H and Harrington R. 2007. Aphids as Crop Pests. CABI North American Office, Cambridge, USA. 699 pp.

Vandermeer JH. 1989. The ecology of Intercropping. Cambridge University Press, Cambridge, UK, 237 pp.

Zhang J, Huang W, Yuan C, Lu Y, Yang B, Wang CY, Zhang P, Dobens L, Zou Z, Wang C and Ling E. 2017. Prophenoloxidase-mediated Ex Vivo immunity to delay fungal infection after insect ecdysis. *Frontiers in Immunology*, 8: 1445.

6.0 GENERAL DISCUSSION, CONCLUSIONS AND RECOMMENDATIONS

6.1 General discussion

Use of synthetic chemical pesticides in management of insect pests including cowpea aphid is very common in sub-Saharan. However, their use is associated with food safety risks especially in vegetables that are regularly harvested in short intervals (Waddington et al., 2010; Egho and Enujeke, 2012; Mweke et al., 2016). Some of the health risks associated with consumption of pesticide residues include short term and long-term effects like headache, discomforts like nausea, different types of cancers, birth deform ties, infertility, and malfunction of endocrine system (Cecchi et al., 2012; Alavan a et al., 2013). These risks are aggravated with consumption of vegetables since some of them are consumed raw or are cooked for a few minutes which does not break down the pesticides. There is therefore the need to develop and adopt pest management strategies that are sustainable and environment and user friendly. Biopesticides including fungal based products are viable alternatives due to their safety to human, environment, non-target beneficial organisms as well as their compatibility with IPM programs.

This study was undertaken to evaluate performance of entomopathogenic fungal isolates against *A. craccivora* under laboratory and screenhouse conditions, identify the suitable formulation of the fungi as well as their efficacy when used in a cowpea monocrop and cowpea-maize intercrop under field conditions. This investigation was necessitated by the need to come up with suitable alternative pest management approach that can be used to meet the increasing demand for safer products especially in vegetable crops. Good performance of entomopathogenic fungi (EPF) isolates against target pest under laboratory and screenhouse conditions does not guarantee stellar performance when used in the field because the efficacy of the EPFs is affected by prevailing environmental conditions (Roy et al., 2006; Eyheraguibel et al., 2010; Jaronski, 2010). Combination of different strategies in management of crop pests is a more sustainable approach and this explains why intercropping cowpea and maize and application of EPF was evaluated in this study.

In Chapter 3, 23 isolates of *M. anisopliae*, *B. bassiana* and *Isaria* sp were evaluated for their pathogenicity against *A. craccivora* in the laboratory. All the isolates were pathogenic to the aphid though the pathogenicity varied between isolates of the same species and also among the different fungal isolates. Three isolates-ICIPE 62 and ICIPE 41 (*M. anisopliae*) and ICIPE 644 (*B. bassiana*) induced highest mortalities within the shortest time (LT_{50}) (in that order) and were selected for mortality dose-response evaluation. In the dose-dependent mortality evaluation, ICIPE 62 produced the least lethal concentration dose that induced 50% mortality (LC_{50}) and comparison of relative potency showed that ICIPE 62 was 50 times and 500 times more potent than ICIPE 41 and ICIPE 644 respectively. The 3 most promising isolates were evaluated for their ability to produce conidia on aphid cadavers at day 3, 6 and 9 post-treatment. ICIPE 62 produced the highest number of spores across all the days evaluated and the spore production increased from day 3, reaching its peak at day 6 and dropping at day 9 post-treatment. The best performing

111

isolate ICIPE 62 was evaluated for its potential to control *A. craccivora* in screenhouse where oil and aqueous formulations were evaluated. Application of ICIPE 62 conidia formulated as either oil or aqueous resulted in reduced aphid population (negative instantaneous) population growth while application of oil or water without conidia resulted in increased *A. craccivora* population (positive instantaneous population growth). Though there are previous studies that reported pathogenicity of *M. anisopliae* against aphid species including *A. craccivora* (Ekesi et al., 2000; Sahayaraj and Borgio, 2010; Saranya et al., 2010; Bayissa et al., 2016a), pathogenicity of isolate ICIPE 62 against *A. craccivora* has never been reported before. Ekesi et al. (2000) reported higher mortalities of *A. craccivora* from *M. anisopliae* isolates from Nigeria but there has been no follow up on commercialization of these potent isolates. Conversely, ICIPE 62 has been commercialized in Kenya for management of aphid species but not *A. craccivora*. The isolate induced high mortality within the shortest time and produced highest number of spores on dead aphids and was more potent compared to other isolates. These are essential attributes that are considered when evaluating potential EPF for registration as a biopesticide (Leng et al., 2011; Niassy et al., 2012; Mascarin et al., 2013; Mohammadbeigi and Port, 2015). Another criterion used in evaluation of a potential biopesticide for registration as a commercial product is its effect on non-target beneficial organisms. Evaluation of effect of ICIPE 62 on beneficial non-target organisms showed that application of the isolate did not result in decline in numbers and diversity of *A. craccivora* predators and parasitoids (Mweke et al., 2017). While Bayissa et al., 2016b reported that ICIPE 62 caused low mortalities on *Cheilomenes lunata* (Fabricus; Coleoptera: Coccinellidae) under laboratory conditions. The information generated in this chapter has identified ICIPE 62 as a potential biopesticide that can be used in the management of aphids in vegetables.

In Chapter 4, the performance of ICIPE 62 formulated as oil and aqueous against *A. craccivora* under field conditions was evaluated for 2 seasons and compared with a synthetic contact insecticide-Lambda-cyhalothrin (Duduthrin® 1.75EC). In the first season characterized by heavy and frequent rainfall and late infestation of *A. craccivora*, the treatments did not result in decrease in aphid population after 8 weeks of application. This observation was attributed to the prevailing weather conditions. It has been reported in previous studies that persistence of EPF based biopesticides after application is critical in initiating an infection that could eventually result in reduced pest population and these are influenced positively or negatively by environmental conditions including rainfall (Inglis et al., 2000; Wraight and Ramos, 2002). Heavy and frequent rainfall washes off conidia on leaf surfaces thereby reducing the number of infective spores available and further reducing the chances of an infection (Inyang et al., 2000). At the same time rainfall has been reported to reduce susceptibility of pests to EPF (Inyang et al., 2000). Distribution of conidia on leaf surface after application has been shown to be influenced by formulation type (Bateman et al., 1993; Bateman and Alves, 2000; Wraight et al., 2002, Santi et al., 2011, Wraight et al., 2016). EPF conidia formulated as oil is reported to result in uniform distribution of the infective spores on leaf surfaces that increases chances of infection, however, this was not observed in the present study. The inability of the EPF formulations to reduce aphid populations could also be attributed to the slow acting nature of the EPF which require incubation period after application before infection could start. Application of Duduthrin did not did not suppress aphid population in the first season though the EPF based formulations did not perform better either:This could be attributed to either heavy rainfall which could have washed off or diluted the pesticide after application or the possibility of a resistant *A. craccivora* population within the experimental site Aphid populations have been reported to comprise of different biotypes or clones with varied resistance to insecticides (Sorensen, 2009; Abdallah et al., 2016).

In the second season which recorded lower and infrequent rainfall, and early and heavy aphid infestations, application of both formulations of the EPF reduced aphid population when compared to control and Duduthrin. This emphasizes the potential of the isolate for use in management of *A. craccivora* under field conditions.

Aphid damages cowpea directly by removing plant sap and indirectly through transmission of viruses that reduce the yield. Honeydew produced by the aphids and deposited on leaves also contributes to yield loss by reducing photosynthetic efficiency and also leaf quality for human consumption. Application of the treatments in both seasons did not prevent cowpea aphid damage compared to control and Duduthrin performed poorer compared to untreated control in the first season. This could be explained by the fact that the cowpea landrace used in this study is highly susceptible to *A. craccivora* and damage as result of aphid population build on susceptible varieties occurs fast (Omoigui et al., 2017). The nature of aphid reproduction results in rapid population build up during favourable weather conditions occasioning damage (Schreiner 2000; Omoigui et al., 2017). In season two both EPF formulations equally protected the plants from aphid damage compared to other treatments. This can partly be attributed to the fact that infection by EPFs reduces feeding and damage by insects especially at terminal stages after infection and before death (Roy et al., 2006).

The two EPF formulations did not negatively affect beneficial arthropods that included ladybird beetles, Spiders (*Leucocage decoratc*), lacewing and parasitoid (*Aphidius colemani*) where else Duduthrin recorded lower beneficial arthropods population. Even though *M. anisopliae* has wide host range, it has been shown to have minimal detrimental effect on natural enemies because some isolates are host specific (Bidochka and Small, 2005; Zimmermann, 2007; Bayissa et al., 2016b).

In the first season characterized by heavy and frequent rainfall with cooler temperatures, both leaf and grain yield were not influenced by treatments because of late infestation by the aphids when leaf yield data collection was almost complete while pod development was at an advanced stage and aphid damage could did not affect the yield. However, in the second season, both EPF formulations produced twice the leaf yield 23.2 (aqueous) and 24.4 (oil) kg ha^{-1}) respectively compared to untreated control (10.5) and Duduthrin 10.9 Kg ha^{-1} respectively. Cowpea grain yield was not influenced by the treatments in the second season and was slightly lower than season one.

This could be explained by the fact that heavy infestations at early stage reduces plant vigour and results in delayed flowering thus reducing harvestable grain yield (Ofuya 1995, 1997).

Chapter 5. The current study demonstrated the efficacy of combining intercropping and application of EPF in suppressing *A. craccivora* population under field conditions. The research findings also demonstrated the impact of environmental conditions on the performance of EPF based biopesticides. The objective in this chapter was to evaluate performance of EPF based biopesticide *M. anisopliae* isolate ICIPE 62 and compare its efficacy with a commonly used pesticide Duduthrin (Lambda-cyhalothrin) in controlling *A. craccivora* in cowpea monocrops and cowpea-maize intercrops.

In the present study intercropping cowpea and maize without application of EPF and application of Duduthrin (a synthetic pyrethroid) in cowpea–maize intercrop did not reduce aphid infestations and damage in all the three seasons. Application of Duduthrin (Lambda-cyhalothrin) which is a commercial insecticide used for aphid control did not suppress *A. craccivora* population either in the monocrop or intercrop. There are similar previous reports where Duduthrin was not able to suppress aphid population after application (Mweke et al., unpublished data, Bayissa, personal communication). This could have been attributed to either heavy rainfall in the first season or existence of a resistant population within the experimental site and its environs. There are previous reports of development of resistance by aphids (Van Emden and Harrington, 2007; Simon, 2011; Abdallah et al., 2016) and although resistance to Duduthrin was not evaluated in this study, this could be a possibility. The efficacy of EPF in controlling aphid under an intercrop system could be attributed to several factors. Intercropping raises relative humidity and the lower crop also benefits from reduced light penetration (Kyamanywa and Ampofo, 1988; Terao, 1997) and these combined can enhance performance of EPF since performance of EPF is impacted by light and relative humidity (Hajek and St. Leger 1994; Jaronski 2010). Intercropping is one of the strategies employed in insect pest management including aphids (Ogenga-Latigo et al. 1992; Hassan 2013; Kisetu et al. 2014). However, studies including the present one have demonstrated that use of this strategy alone does not guarantee successful pest management. In this study, application of the either the EPF or Duduthrin did not negatively impact cowpea aphid natural enemies abundance and diversity, however, the parasitoid (*Aphidius colemani*) recorded the lowest numbers in all the 3 seasons. The parasitoid population under natural environment is affected by biotic and abiotic factors like temperature, relative humidity, presence of other beneficial arthropods like predators and pesticide residues in the environment (Brodeur and Rosenheim, 2000; Bilu and Coll, 2007; Colinet et al., 2007; Araya et al., 2010; Prado et al., 2015) These could have played a role in the lowering the *A. colemani* population in the experimental site.

This study has demonstrated the potential of use of EPF in suppressing aphid population and conferring yield benefits to farmers. In the second season aphid infestations occurred early and application of EPF in an intercrop system was able to protect the crop from aphid damage compared to untreated intercrop and an intercrop treated with Duduthrin.

Application of EPF based biopesticides in cowpea-maize intercrop systems can be suitable alternatives to use of synthetic chemical pesticides and could therefore provide a more sustainable management strategy for aphids since the biopesticides are user friendly, safe to environment and do not harm natural enemies of aphids. The *M. anisopliae* isolate used in this study (ICIPE 62) has known pathogenicity to several aphid species including *A. craccivora* (Bayissa et al., 2016a; Mweke et al., 2018) and has been developed by *icipe* and commercialized by Real IPM for use in management of aphids (http://www.realipm.com/).

6.2 Conclusions and recommendations

This study has identified a virulent *Metarhizium anisopliae*-isolate (ICIPE 62) that can be used in management of cowpea aphid, *Aphis craccivora* in cowpea and other vegetables under field conditions. The results presented here have also demonstrated the impact of environmental conditions on performance of EPF based biopesticide under field conditions. However, biopesticides offer promising alternatives to chemical pesticides and are suitable for vegetables that have short harvesting intervals and some of which are consumed without being cooked as they do not leave toxic residues. The research findings have has also demonstrated the efficacy of an integrated strategy for management of cowpea aphid through application of entomopathogenic fungi in a cowpea- maize intercrop. This approach is acceptable to smallholder farmers who grow cowpea as an intercrop with cereals and who constitute the majority of cowpea producers in Kenya and therefore could easily adopt the management approach.

The study recommends that further evaluation of the effect of the EPF on other natural enemies of aphids, that were not evaluated, should be undertaken. It further recommends that more research needs to be carried out to enhance formulations of the EPF based biopesticides to improve their persistence after application and reduce undesirable effects of environmental conditions, while enhancing their performance under different weather conditions. Studies should be undertaken on different additives, adjuvants and protectants that improve EPF based formulations to improve and enhance their performance under different environmental conditions. Investigations on how *M. anisopliae* ICIPE 62 affected the biology of *A. craccivora* resulting in reduction of its population needs to be carried out. Finally, it is recommended that the label for ICIPE 62, which is commercially registered for management of the aphid species: *Brevicoryne brassicae*, *Lipaphis pseudobrassicae* and *Aphis gossypii*, should be expanded to include *A. craccivora*. This would cut the period required for evaluation of the biopesticide and the associated costs and would be a relief to the company that is already selling the biopesticide.

Schlussfolgerungen und Empfehlungen

Die vorliegende Studie hat ein virulentes Isolat von *Metarhizium anisopliae* (ICIPE 62) identifiziert, das zur Bekämpfung der Schwarzen Bohnenblattlaus, *Aphis craccivora*, im Anbau von Bohnen und anderen Gemüsen unter Freilandbedingungen eingesetzt werden kann. Die Ergebnisse haben gezeigt, dass die optimale Leistung von Biopestiziden auf Basis von entomopathogenen Pilzen stark von den Umweltbedingungen abhängt. Biopestizide bieten jedoch vielversprechende Alternativen zu chemischen Pestiziden und eignen sich für Gemüse mit kurzen Ernteintervallen, von denen einige ohne gekocht zu werden verzehrt werden, da sie keine toxischen Rückstände hinterlassen. Im Rahmen der Forschungen wurde der erfolgreiche Einsatz des entomopathogenen Pilzes im Rahmen einer integrierten Strategie für das Management der Schwarzen Bohnenblattlaus in einem Zwischenanbau von Bohnen und Mais nachgewiesen. Dieser Ansatz ist für Kleinbauern akzeptabel, die Bohnen und Zwischenfrüchte mit Getreide anbauen. Hierzu zählen in Kenia die Mehrheit der Kuhbohnenproduzenten.

Die Studie empfiehlt, die Wirkung des eingesetzten entomopathogenen Pilzes auf andere natürliche Blattlausfeinde, die bisher nicht untersucht wurden, weiter zu untersuchen. Es wird ferner empfohlen, mehr Forschung zu betreiben, um die Formulierungen der entomopathogenen Pilze zu verbessern. Hier geht es in erster Linie um die Persistenz in der Umwelt nach der Anwendung sowie die Abhängigkeit der Wirksamkeit von Umweltbedingungen. Es müssen Untersuchungen durchgeführt werden, wie *M. anisopliae* ICIPE 62 die Biologie von *A. craccivora* beeinflusst, welches letztendlich zu einer Verringerung der Population führt. Schließlich wird empfohlen, dass die Zulassung von ICIPE 62, welches für das Management der folgenden Blattlausarten kommerziell registriert ist: *Brevicoryne brassicae*, *Lipaphis pseudobrassicae* und *Aphis gossypii* um die Art *A. craccivora* zu erweitern. Dies würde den Zeitaufwand für die Bewertung des Biopestizids und der damit verbundenen Kosten im Rahmen der Zulassung reduzieren und wäre eine Entlastung für die Wirtschaft.

6.3 References

Abdallah IS, Abou-Yousef HM, Fouad EA and Kandil MAEH. 2016. The role of detoxifying enzymes in the resistance of the cowpea aphid (*Aphis craccivora* Koch) to thiamethoxam. *Journal of Plant Protection Research*, 56 (1):67-72.

Alavanja MC, Ross MK and Bonner MR. 2013. Increased cancer burden among pesticide applicators and others due to pesticide exposure. CA: *A Cancer Journal for Clinicians*, 63(2):120-142.

Araya JE, Araya M and Guerrero MA. 2010. Effects of some insecticides applied in sublethal concentrations on the survival and longevity of *Aphidius ervi* (Haliday) (Hymenoptera: Aphidiidae) Adults. *Chilean Journal of Agricultural Research*, 70 (2):221-227.

Bateman R, Carey M, Moore D and Prior C. 1993. The enhanced infectivity of *Metarhizium flavoviride* in oil formulations to desert locusts at low humidities. *Annals of Applied Biology*, 122:145-152.

Bateman RP and Alves RT. 2000. Delivery systems for mycoinsecticides using oil-based formulations. *Aspects of Applied Biology*, 57:163-170.

Bayissa W, Ekesi S, Mohamed SA, Kaaya GP, Wagacha JM, Hanna R and Maniania NK. 2016a. Selection of fungal isolates for virulence against three aphid pest species of crucifers and okra. *Journal of Pest Science*, 90:355-68.

Bayissa W, Ekesi S, Mohamed SA, Kaaya GP, Wagacha JM, Hanna R and Maniania NK. 2016b. Interactions among vegetable-infesting aphids, the fungal pathogen *Metarhizium anisopliae* (Ascomycota: Hypocreales) and the predatory coccinellid *Cheilomenes lunata* (Coleoptera: Coccinellidae). *Biocontrol Science and Technology*, 26:274-290.

Bidochka MJ and Small CS. 2005. Phylogeography of *Metarhizium,* an insect pathogenic fungus. In Vega FE and Blackwell M (Eds.), Insect-fungal associations: Ecology and evolution. New York, NY: Oxford University Press, 28–50 pp.

Bilu E and Coll M. 2007. The importance of intraguild interactions to the combined effect of a parasitoid and a predator on aphid population suppression. *BioControl*, 52:753–763.

Brodeur J and Rosenheim JA. 2000. Intraguild interactions in aphid parasitoids. *Entomologia Experimentalis et Applicata*, 97(1):93-108.

Cecchi A; Rovedatti GM, Sabino G and Magnarelli G. 2012. Environmental exposure to organophosphate pesticides: Assessment of endocrine disruption and hepatotoxicity in pregnant women. *Ecotoxicology and Environmental Safety*, 80:280–287.

Egho EO and Enujeke EC. 2012. Minimizing insecticide application in the control of insect pests of cowpea (*Vigna unguiculata* (L) WALP) in Delta state, Nigeria. *Sustainable Agriculture Research*, 1 (1):87.

Ekesi S, Akpa AD, Onu I and Ogunlana MO. 2000. Entomopathogenicity of *Beauveria bassiana* and *Metarhizium anisopliae* to the cowpea aphid, *Aphis craccivora* Koch (Homoptera: Aphididae). *Archieves of Phytopathology and Plant Protection*, 33 (2):171–180.

Eyheraguibel B, Richard C, Ledoigt G and Ter Halle A. 2010. Photoprotection by plant extracts: a new ecological means to reduce pesticide photodegradation. *Journal of Agriculture and Food Chemistry*, 58 (17):9692-9696.

Hajek AE, St. Leger RJ. 1994. Interactions between fungal pathogens and insect hosts. *Annual Review of Entomology*, 39 (1):293–322.

Hassan S. 2013. Effect of variety and intercropping on two major cowpea (*Vigna unguiculata* L. Walp) field pests in Mubi, Adamawa state, Nigeria. *International Journal of Agricultural Research and Development*, 1 (5):108-109.

Inglis GD, Ivie TJ, Duke GM and Goettel MS. 2000. Influence of rain and conidial formulations on persistence of *Beauveria bassiana* on potato leaves and Colorado potato beetle larvae, *Biological Control*, 18 (1):55-64.

Inyang EN, McCartney HA, Oyejola B, Ibrahim L, Pye BJ, Archer S and Butt TM. 2000. Effect of formulation, application and rain on the persistence of the entomogenous fungus *Metarhizium anisopliae* on oilseed rape. *Mycological Research*, 104 (6):653-661.

Jaronski, ST. 2010. Ecological factors in the inundative use of fungal entomopathogens. *BioControl*, 55 (1):159-185.

Kisetu E, Nyasasi BT and Nyika M. 2014. Effect of cropping systems on infestation and severity of field pests of cowpea in Morogoro, Tanzania. *Modern Journal of Agriculture*, 1(1):1-9.

Kyamanywa S and Ampofo JKO. 1988. Effect of cowpea/maize mixed cropping on the incident light at the cowpea canopy and flower thrips (Thysanoptera: Thripidae) population density. *Crop Protection*, 7 (3):186-189.

Lopes RB, Pauli G, Mascarin GM and Faria M. 2011. Protection of entomopathogenic conidia against chemical fungicides afforded by an oil-based formulation. Biocontrol Science and Technology, 21 (2):125-137. Mascarin GM, Kobori NN, Quintela ED and Delalibera I. 2013. The virulence of entomopathogenic fungi against *Bemisia tabaci* biotype B (Hemiptera: Aleyrodidae) and their conidial production using solid substrate fermentation. *Biological control*, 66(3):209-218.

Mohammadbeigi A and Port G. 2015. Effect of infection by *Beauveria bassiana* and *Metarhizium anisopliae* on the feeding of *Uvarovistia zebra*. *Journal of Insect Science*, 15(1):88.

Mweke A, Ekesi S, Maniania NK, Fiaboe KKM, Ulrichs C. 2017. Performance of *Metarhizium anisopliae* (Metsch.) Sorok and *Beauveria bassiana* (Bals.) Vuill. isolates against cowpea aphid (*Aphis craccivora* Koch) in cowpea under field conditions, Tropentag Conference, 20th -23rd September, Bonn, Germany.

Mweke A, Ulrichs C, Maniania KN and Ekesi S. 2016. Integration of entomopathogenic fungi as biopesticide for the management of cowpea aphid (*Aphis craccivora* Koch). *African Journal of Horticultural Science*. 9:14–31.

Mweke A, Ulrichs C, Nana P, Akutse KS, Fiaboe KKM, Maniania NK and Ekesi S. 2018. Evaluation of the entomopathogenic fungi *Metarhizium anisopliae, Beauveria bassiana* and *Isaria* sp. for the management of *Aphis craccivora*. *Journal of Economic Entomology*, 111 (4):2018, 1587–1594.

Niassy S, Maniania N, Subramanian S, Gitonga L, Mburu D, Masiga D and Ekesi S. 2012. Selection of promising fungal biological control agent of the western flower thrips *Frankliniella occidentalis* (Pergande). *Letters in Applied Microbiology*, 54 (6):487-493.

Ofuya TI. 1995. Studies on the capability of *Cheilomenes lunata* (Fabricius) (Coleoptera: Coccinellidae) to prey on the cowpea aphid, *Aphis craccivora* Koch Homoptera: Aphididae) in Nigeria. *Agriculture Ecosystems and Environment*, 52 (1):35-38.

Ofuya TI. 1997. Control of the bean aphid *Aphis craccivora* Koch (Homoptera: Aphididae) in cowpea, *Vigna unguiculata* (L.) Walp. *Integrated Pest Management Reviews* 2 (4):199-207.

Ogenga-Latigo MW, Ampofo JKO and Balidawa CW. 1992. Influence of maize row spacing on infestation and damage of intercropped beans by bean aphid (*Aphis fabae* Scop.) I. Incidence of Aphids. *Field Crop Research*, 30 (1-2):111-121.

Omoigui LO, Ekeuro GC, Kamara AY, Bello LL, Timko MP and Ogunwolu GO. 2017. New sources of aphids (*Aphis craccivora* (Koch) resistance in cowpea germplasm using phenotypic and molecular marker approaches. *Euphytica*, 213:178.

Prado SG, Jandricic SE and Frank SD. 2015. Ecological interactions affecting the efficacy of *Aphidius colemani* in greenhouse crops. *Insects*, 6(2):538-575.

Roy HE, Steinkraus DC, Eilenberg J, Hajek AE and Pell JK. 2006. Bizarre interactions and endgames: entomopathogenic fungi and their arthropod hosts. *Annual Review of Entomology*, 51:331-357.

Sahayaraj K and Borgio JF. 2010. Virulence of entomopathogenic fungus *Metarhizium anisopliae* (Metsch.) Sorokin on seven insect pests. *Indian Journal of Agricultural Research*, 44 (3):195-200.

Saidi M, Itulya FM, Aguyoh JN, Mshenga PM, Owour G. 2010b. Effects of cowpea leaf harvesting initiation time on yields and profitability of a dual-purpose sole cowpea and cowpea-maize intercrop. *Electronic Journal of Environmental, Agricultural and Food Chemistry*, 9(6): 1134-1144.

Santi LE, Silva LA, da Silva WOB, Corrêa APF, Rangel DEN, Carlini CR, Schrank A and Vainstein MH. 2011. Virulence of the entomopathogenic fungus *Metarhizium anisopliae* using soybean oil formulation for control of the cotton stainer bug, *Dysdercus peruvianus*. *World Journal of Microbiology and Biotechnology*, 27 (10):2297-2303.

Schreiner I. 2000. Cowpea Aphids, Agricultural pests of the pacific. ADAP 2000-6, Reissued February 2000. ISBN 1-931435-09-X.

Simon JYu. 2011. The Toxicology and Biochemistry of Insecticides. CRC Press, Taylor & Francis Group, Boca Raton, USA, 211 pp.

Sorensen JT. 2009. Aphids. In Encyclopedia of Insects. Resh, V.H. and Cardé, R.T. (eds)

Terao T, Watanabe I, Matsunaga R, Hakoyama S and Singh BB. 1997. Agro-physiological constraints in intercropped cowpea: an analysis: In B. B. Singh, D. R. Mohan Raj, K. E. Dashiell, and L.E.N. Jackai [eds.], Advance in cowpea research. Sayce, Devon, UK, 129-140 pp.

Van Emden H, Harrington R. 2007. Aphids as Crop Pests. CABI North American Office, Cambridge, USA, 699 pp.

Waddington SR, Li X, Dixon J, Hyman G and De Vicente MC. 2010. Getting the focus right: production constraints for six major food crops in Asian and African farming systems. *Food Security*, 2 (1):27-48.

Wraight SP and Ramos ME. 2002. Application parameters affecting field efficacy of *Beauveria bassiana* foliar treatments against Colorado potato beetle *Leptinotarsa decemlineata*. *Biological Control*, 23 (2):164-178.

Wraight SP, Filotas MJ and Sanderson JP. 2016. Comparative efficacy of oil-oil, wettable-powder, and unformulated-powder preparations of *Beauveria bassiana* against the melon aphid *Aphis gossypii*, *Biocontrol Science and Technology*, 26 (7):894-914.

Zimmermann G. 2007. Review on safety of the entomopathogenic fungus *Metarhizium anisopliae*. *Biocontrol Science and Technology*, 17 (9):879–920

ACKNOWLEDGEMENTS

The research was carried out with financial support from the German Federal Ministry of Economic Cooperation and Development (BMZ) and the German Federal Ministry of Education and Research (BMBF). We gratefully acknowledge aid from the UK Government; Swedish International Development Cooperation Agency (Sida); the Swiss Agency for Development and Cooperation (SDC) and the Kenyan Government through their core financial support to icipe research portfolio. I would like to appreciate the BMZ funded Dissertation Research Internship Programme (DRIP) of icipe for financing my studies.

I am extremely grateful to Professor Dr. Dr. Christian Ulrichs for his unwavering support advice, and guidance throughout my PhD research. My appreciation also goes to Dr. Sunday Ekesi for guiding me through proposal development, logistical and intellectual support and guidance during execution of the research proposal and write up. Dr. Komi Fiaboe, Dr. Komivi Senyo Akutse, Dr. Nguya Kalemba Maniania and Dr. Paulin Nana are appreciated for their commitment and support during data collection and write up of manuscripts included in this thesis. I would also like to thank National Museums of Kenya and Kenya Plant Health Inspectorate Service (KEPHIS) for their assistance in identifying some of the insects collected during my PhD field study. My gratitude also goes to Dr. Daisy Salifu and Mr. Benedict Orindi for their statistical advice and my appreciations to Ms. Jane Kimemia, Mr. Levi Ombura, Mr. Joshua Mutuku, Mr Joseck Esikuru and Mr. Pascal Agola for their technical assistance during my study. I am also extremely grateful to my family for their unwavering support throughout my studies.

In der Reihe *Berliner ökophysiologische und phytomedizinische Schriften* **sind bisher erschienen:**

Band 01:   Mohammad Mahir Uddin (2009)
            Chemical ecology of mustard leaf beetle Phaedon cochleariae (F.).
            ISBN 978-3-89959-848-3.

Band 02:   Ilir Morina (2009)
            Entwicklung von Verfahren zur Rekultivierung der Aschedeponie des
            Braunkohlekraftwerks in Prishtina (Kosovo).
            ISBN 978-3-89959-872-8.

Band 03:   Melanie Wiesner (2009)
            Veränderungen gesundheitsrelevanter Inhaltsstoffe in *Parthenium
            hysterophorus* L. in Abhängigkeit von der Pflanzengröße und Klimafaktoren.
            ISBN 978-3-89959-880-3.

Band 04:   Fransika Rohr (2009)
            Variabilität aliphatischer Glucosinolate in *Arabidopsis thaliana*-Ökotypen und
            deren Einfluss auf die Wirtspflanzeneignung von zwei folivoren Insektenarten.
            ISBN 978-3-89959-884-9.

Band 05:   Jutta Buchhop (2009)
            Characterization of phylogenetically diverse CLRV-isolates by RFLP and
            research into identification of two isometric viruses.
            ISBN 978-3-89959-929-9.

Band 06:   Nora Koim (2010)
            Urban sprawl, land cover change and forest fragmentation – Case study
            Pereira, Colombia.
            ISBN 978-3-89959-955-8.

Band 07:   Nadja Förster (2010)
            Eignung unterschiedlicher salicylathaltiger *Salix*-Klone für die
            Arzneimittelindustrie.
            ISBN 978-3-89959-964-0.

Band 08:   Jana Gentkow (2010)
            Cherry leaf roll virus (CLRV): Charakterisierung ausgewählter Virusisolate
            unter besonderer Berücksichtigung des viralen Hüllproteins.
            ISBN 978-3-89959-976-3.

Band 09:   Ahmad Fakhro (2010)
            Interaction of Pepino mosaic virus (PepMV) and fungal root endophytes with
            tomato hosts (*Lycopersicum esculentum* Mill.).
            ISBN 978-3-89959-995-4.

Band 10:   Stefan Irrgang (2010)
            Mikro- und makroskopische Untersuchungen an Veredelungsstellen von
            Straßenbäumen im Hinblick auf die Beeinflussung ihrer Bruchsicherheit.
            ISBN 978-3-89959-998-5.

Band 11:   Julia Jahnke (2010)
            Guerilla Gardening anhand von Beispielen in New York, London und Berlin.
            ISBN 978-3-86247-001-3.

Band 12:   Astrid Karoline Günther (2010)
Analysen zur Intensität der Pflanzenschutzmittel-Anwendung und Aufklärung
ihrer Einflussfaktoren in ausgewählten Ackerbaubetrieben.
ISBN 978-3-85247-005-1.

Band 13:   Milena A. Dimova (2010)
Untersuchungen zur Epidemiologie von *Pythium aphanidermatum* in
Abhängigkeit von den Umgebungsbedingungen bei der Gewächshausgurke
(*Cucumis sativus* L.).
ISBN 978-3-86247-033-4.

Band 14:   Claudia Patricia Pérez-Rodríguez (2010)
Physiologische Veränderungen in Früchten der Solanaceaengewächse in
Abhängigkeit von physikalischen Elicitoren während der Produktion und nach
der Ernte.
ISBN 978-3-86247-066-2.

Band 15:   Charles Adarkwah (2010)
Integrated management of the stored-product pest insects *Corcyra cephalonica*,
Cadra cautella, *Sitophilus zeamais* and *Tribolium castaneum* by use of the
parasitic wasps *Habrobracon hebetor*, *Venturia canescens*, *Lariophagus
distinguendus* and neem seed oil.
ISBN 978-3-86247-077-8.

Band 16:   Christoph von Studzinski (2010)
Angewandte Methoden der xenovegetativen Vermehrung.
ISBN 978-3-86247-088-4.

Band 17:   Tanja Mucha-Pelzer (2011)
Amorphe Silikate – Möglichkeiten des Einsatzes im Gartenbau zur
physikalischen Schädlingsbekämpfung.
ISBN 978- 3-86247-106-5.

Band 18:   Diego Miranda (2011)
Effect of salt stress on physiological parameters of cape gooseberry, *Physalis
peruviana* L.
ISBN 978- 3-86247-119-5

Band 19:   Franziska Beran (2011)
Host preference and aggregation behavior of the striped flea beetle, *Phyllotreta
striolata*.
ISBN 978- 3-86247-188-1

Band 20:   Mohammed Abul Monjur Khan (2011)
Induced biochemical changes and gene expression in *Brassica oleracea* and
Arabidopsis thaliana by drought stress and its consequences on resistance to
aphids.
ISBN 978- 3-86247-203-1.

Band 21:   Sandra Lerche (2012)
Untersuchungen zur Anwendung, Praxiseinführung und molekularen
Identifizierung von Stamm V24 des entomopathogenen Pilzes *Lecanicillium
muscarium* (Petch) Zare & W. Gams.
ISBN 978- 3-86247-248-2.

Band 22:   Carsten Richter (2012)
             Entwicklung und Überprüfung eines gasdichten Küvettensystems für
             Experimente unter hochgradig kontrollierten Bedingungen mit
             Gaswechselmessungen.
             ISBN 978- 3-86247-271-0.

Band 23:   Aksana Grineva (2012)
             Influence of the two stored grain pest insects *Sitophilus granarius* and
             Oryzaephilus surinamensis on temperature, relative humidity, moisture
             content, and mould growth in stored triticale.
             ISBN 978- 3-86247-279-6.

Band 24:   Carmen Büttner & Christian Ulrichs (2012)
             Aktuelle Themen in Landwirtschaft und Gartenbau am Beispiel von Südtirol.
             ISBN 978- 3-86247-279-6.

Band 25:   Juliane Langer (2012)
             Molecular and epidemiological characterisation of Cherry leaf roll virus
             (CLRV).
             ISBN 978- 3-86247-279-6.

Band 26:   Franziska Rohr-Doucet (2012)
             AOP-Variabilität in *Arabidopsis thaliana*-Kreuzungslinien – Auswirkungen
             auf die Resistenz gegenüber verschieden spezialisierten Lepidopteren-Arten.
             ISBN 978- 3-86247-329-8.

Band 27:   Vanessa Hörmann (2012)
             Lignin als biologische Barriere gegen Schimmelpize in Innenräumen.
             ISBN 978- 3-86247-330-4.

Band 28:   Jacqueline Kurth (2013)
             Auswirkungen verschiedener Düngerzusammensetzungen auf den Ertrag bei
             Schnittrosen unter Berücksichtigung des Anbauverfahrens.
             ISBN 978- 3-86247-336-6.

Band 29:   Juliane Langer, Carmen Büttner & Christian Ulrichs (2014)
             Kolumbien – klimatische und politische Voraussetzungen für eine
             landwirtschaftliche Produktion.
             ISBN 978- 3-86247-430-1.

Band 30:   Heike Luisa Dieckmann (2014)
             Detection of the European mountain ash ringspot associated virus (EMARaV)
             in Sorbus aucuparia L. in several European contries.
             ISBN 978- 3-86247-441-7.

Band 31:   Rima Marion Baag (2014)
             Analyse von trans-Resveratrol in historischen Rebsorten der Weinanbaugebiete
             Sachsen und Saale-Unstrut.
             ISBN 978- 3-86247-488-2.

Band 32:   Ayesha Rahmann (2014)
             Study of the protective effects of nano-structured silica and plant derived
             biomolecules on nuclear polyhedrosis virus affected silkworm larvae at the
             behavioral and molecular level.
             ISBN 978- 3-86247-495-0.

Band 33:   Bettina Gramberg (2015)
           Weiterentwicklung eines elektrochemischen Biosensors zum Nachweis von
           Pflanzenviren und Insektiziden.
           ISBN 978- 3-86247-512-4.

Band 34:   Wilhelm van Husen (2015)
           Artspezifische Aufnahme und Verteilung von Cadmium bei indigenen
           afrikanischen Gemüsearten und daraus abzuleitende Ernährungsempfehlungen.
           ISBN 978- 3-86247-523-0.

Band 35:   Jenny Roßbach (2015)
           European mountain ash ringspot-associated virus (EMARaV): diversity and
           geographic distribution in Europe.
           ISBN 978- 3-86247-547-6.

Band 36:   Silke Steinmöller (2015)
           Risikominderung der Verbreitung von Quarantäneschadorganismen der
           Kartoffel durch hygienisierende Maßnahmen.
           ISBN 978- 3-86247-550-6.

Band 37:   Christin Siewert (2016)
           Genomic and functional analysis of species within the Acholeplasmataceae –
           Phytoplasmas and Acholeplasmas.
           ISBN 978- 3-86247-579-7.

Band 38:   Angela Köhler (2016)
           Untersuchungen zur Phenolglycosidkonzentration ausgewählter intra- und
           interspezifischer Kreuzungen salicinreicher Biomasseweiden.
           ISBN 978- 3-86247-581-0.

Band 39:   Nicolas Meyer (2016)
           Vergleichende ökophysiologische Untersuchung verschiedener Baumarten zur
           Verwendung als Straßenbegleitgrün in Berlin.
           ISBN 978- 3-86247-586-5.

Band 40:   Stefanie Schläger (2017)
           Identification of variation within sex pheromone blends of various Maruca
           vitrata populations for refining pheromone lures and traps in Asia.
           ISBN 978- 3-7369-9570-3.

Band 41:   Elisha Otieno Gogo (2017)
           Pre- and postharvest treatments for the quality assurance of African indigenous
           leafy vegetables.
           ISBN 978-3-7369-9650-2.

Band 42:   Luise Dierker (2017)
           Interaktion des RNA2-kodierten Transportproteins (MP) des Cherry leaf roll
           virus (CLRV) mit dem viralen Hüllprotein (CP) und pflanzlichen
           Wirtsfaktoren
           ISBN 978-3-7369-9670-0.

Band 43:   Nadja Förster (2017)
           Antikarzinogenes Potential ausgewählter Glucosinolate von Moringa oleifera
           ISBN 978-3-7369-9704-2.

Band 44:   Steffen Pallarz (2018)
             Data driven classification of host-plant response (virus-plant)
             ISBN 978-3-7369-9731-8.

Band 45:   Vanessa Hörmann (2018)
             Biofiltration of indoor pollutants by ornamental plants
             ISBN 978-3-7369-9815-5.

www.ingramcontent.com/pod-product-compliance
Lightning Source LLC
Chambersburg PA
CBHW061333220326
41599CB00026B/5172